# Geology of the country around Chester and Winsford

The district described in this memoir lies on the western edge of the Cheshire Basin, a major pull-apart basin during Triassic times in which sedimentation was controlled by large N-trending faults. The oldest rocks coming to crop in the district are Upper Coal Measures but their outcrop is entirely covered in drift. Their sequence is proved, together with that of the underlying Carboniferous and Ordovician rocks, in deep boreholes sunk in the search for coal and oil.

Hills of Sherwood Sandstone range northwards through the middle of the district and afford important sections. Together with sequences proved in several cored boreholes, these allow the overall picture of Triassic sedimentation to be built up. Fossils are rare in these predominantly non-marine rocks but studies of the fossil spores have established a broad correlation between this local succession and the standard Triassic stages.

Beds of salt beneath the Winsford-Northwich area have long been worked and a detailed account is given of their stratigraphy and structure. The history of local salt mining is described and includes the disastrous subsidences caused by brine pumping. The unstable ground of the salt-fields has implications for building and construction even at the present day.

Much of the district is covered by thick glacial deposits of Devensian age. These are described in detail and their stratigraphical relationships are emphasised.

Other sections of the memoir deal with the tectonic history, post-glacial deposits, mineral products and hydrogeology of the district.

**Frontispiece**  Bottom Flash, a widening of the River Weaver where it crosses an area of subsidence caused by brine pumping at Winsford (in the distance)

BRITISH GEOLOGICAL SURVEY

J. R. EARP and
B. J. TAYLOR

CONTRIBUTORS

*Stratigraphy*
D. Magraw and
E. G. Poole

*Palaeontology*
M. Mitchell, S. G. Molyneux,
N. J. Riley and G. Warrington

*Petrology*
R. K. Harrison

*Hydrogeology*
R. Ireland

# Geology of the country around Chester and Winsford

Memoir for 1:50 000 geological sheet 109

Natural Environment Research Council

LONDON: HER MAJESTY'S STATIONERY OFFICE   1986

*First published 1986*

ISBN 0 11 884392 3

*Bibliographical reference*

EARP, J. R. and TAYLOR, B. J. 1986. Geology of the country around Chester and Winsford. *Mem. Br. Geol. Surv.*, Sheet 109.

*Authors*

J. R. Earp, MSc, PhD
*South Chiltern, Orchard Coombe, Whitchurch Hill, Berkshire RG8 7QL*

B. J. Taylor, BSc
*Fiddlers Green, 10 Quietways, Stonehouse, Gloucester GL10 2NW*

*Contributors*

The late D. M. Magraw, DSc

E. G. Poole, BSc
*20 Fountains Avenue, Boston Spa, North Yorkshire*

M. Mitchell, MA, S. G. Molyneux, BSc, PhD,
N. J. Riley, BSc, PhD, G. Warrington, BSc, PhD
*British Geological Survey, Keyworth, Nottingham NG12 5GG*

R. K. Harrison, MSc
*27 Springfield Park, Twyford, Berkshire*

R. Ireland, BSc
*North West Water Authority*
*Dawson House, Great Sankey, Warrington WH5 3LW*

*Other publications of the Survey dealing with this district and adjoining districts*

BOOKS

*Memoirs*

Geology of the country around Stockport and Knutsford, Sheet 98
Geology of the country around Macclesfield, Congleton, Crewe and Middlewich, Sheet 110
Geology of the country around Nantwich and Whitchurch, Sheet 122

*British Regional Geology*

The Pennines and adjacent areas (3rd edition)
Central England (3rd edition)
North Wales (3rd edition)

MAPS

*1:625 000*
Solid geology (South sheet)
Quaternary geology (South sheet)
Aeromagnetic map (South sheet)

*1:250 000*
Solid geology, Liverpool Bay
Seabed sediments, Liverpool Bay
Bouguer gravity anomaly, Liverpool Bay
Aeromagnetic anomaly, Liverpool Bay
Mid Wales and Marches

*1:63 360 and 1:50 000 (Solid and Drift)*
Sheet 96      Liverpool
Sheet 97      Runcorn
Sheet 98      Stockport
Sheet 108     Flint
Sheet 109     Chester
Sheet 110     Macclesfield
Sheet 121     Wrexham
Sheet 122     Nantwich

Produced in the United Kingdom for HMSO

Dd 738927   C20   9.86   49202

# CONTENTS

FIGURES

## PLATES

# PREFACE

It is now more than a century since the publication in 1882 of a short Memoir by A. Strahan describing the geology of the Old Series one-inch quarter sheet 80 SW entitled 'Geology of the Neighbourhood of Chester'. Dealing with almost all except the easternmost quarter of the area of the Chester (109) New Series Sheet, the work was a summary of the results of Geological Survey investigations, mainly at one-inch-to-one-mile, during the preceding 30 years or so, first by E. Hull and later by A. Strahan and C. E. De Rance. The contemporary surveys of the adjoining Old Series one-inch quarter sheet 80 SE were never made the subject of a Memoir.

The primary six-inch-to-one-mile survey was carried out in the years 1953 to 1956 by Mr B. J. Taylor, Dr J. R. Earp, Dr D. Magraw and Mr E. G. Poole under the supervision of F. M. Trotter and J. V. Stephens as District Geologists. The initials of the responsible officers are shown on p.viii against a list of the six-inch maps that cover the Sheet. On completion of the surface mapping of the district, the Geological Survey commissioned nine exploratory boreholes sited to provide new information on the solid rocks where they were obscured by thick superficial deposits.

This account by Dr Earp and Mr Taylor incorporates notes provided by Mr Poole and Dr Magraw; contributions of the individual officers are indicated by appended initials. Mr R. K. Harrison has provided notes on the cleaved mudstone and metabasite underlying the Carboniferous rocks at Milton Green. Colleagues in the Biostratigraphy Research Group have provided determinations of fossils from Ordovician, Carboniferous, Triassic and Quaternary sediments. Dr R.A. Old has collated the text and Mr W. B. Evans has edited the entire work.

We gratefully acknowledge the cooperation of Imperial Chemical Industries Ltd., who provided a large amount of mining and borehole information in the saltfield; also that of the Cheshire Brine Subsidence Compensation Board who provided information on old workings and allowed us to publish photographs from their archives. We are indebted to the Shell Refining and Marketing Company Ltd. for supplying borehole information from the Thornton-le-Moors area, to the Esso Petroleum Company for permission to use and publish data from their deep well Milton Green No. 1, and to the National Coal Board for releasing some results from its recent drilling.

Our thanks are due to the many former water undertakings of the rural and urban authorities, and to their successors, the North West Water Authority, for much valuable geological and hydrogeological information. We especially thank Mr R. Ireland of the North West Water Authority for writing the section on hydrogeology. Finally, we express our appreciation of the courtesy and forbearance of the many owners and tenants of land who freely allowed us access to their fields and woodlands to carry out our surveys and to drill boreholes.

G. Innes Lumsden, FRSE
*Director*

*British Geological Survey*
*Keyworth*
*Nottingham    NG12 5GG*

21st May 1986

# LIST OF SIX-INCH MAPS

The following list shows the six-inch maps included wholly or partly within the area of Sheet 109 (Chester) of the 1:50 000 Geological Map of England and Wales together with the initials of the surveyors and the dates of survey. Uncoloured dyeline copies of these maps, except those indicated by an asterisk lying along the western margin of the district, may be purchased from the office of the British Geological Survey at Keyworth, Nottingham. They are available for public reference at that office. The surveyors were J. R. Earp, D. Magraw, E. G. Poole, B. J. Taylor and F. M. Trotter. National Grid maps : all within the 100 kilometre square SJ.

| | | | |
|---|---|---|---|
| 35 NE* | Rossett | EGP | 1956 |
| 36 SE* | Dodleston | DM, EGP | 1953–54 |
| 36 NE* | Blacon | JRE, DM | 1953–54 |
| 37 SE* | Capenhurst | JRE | 1954 |
| 45 NW | Churton | EGP | 1954 |
| 45 NE | Handley | EGP | 1954–55 |
| 46 SW | Eccleston | DM, EGP | 1953–54 |
| 46 SE | Waverton | DM, EGP | 1953–55 |
| 46 NW | Chester | JRE, DM | 1953–55 |
| 46 NE | Tarvin | JRE, DM | 1953–55 |
| 47 SW | Stoak | JRE | 1953–56 |
| 47 SE | Dunham-on-the-Hill | JRE | 1953–56 |
| 55 NW | Peckforton | EGP | 1955 |
| 55 NE | Bunbury | EGP, DM | 1955 |
| 56 SW | Clotton | DM, EGP | 1953–56 |
| 56 SE | Tarporley | DM, BJT, EGP | 1955–56 |
| 56 NW | Kelsall | JRE, DM | 1953–56 |
| 56 NE | Delamere | JRE, DM, BJT, FMT | 1954–56 |
| 57 SW | Manley | JRE, BJT | 1953–56 |
| 57 SE | Norley | BJT, FMT | 1956 |
| 65 NW | Wardle | BJT | 1954–55 |
| 65 NE | Leighton | BJT | 1956 |
| 66 SW | Wettenhall | BJT | 1953–56 |
| 66 SE | Wimboldsley | BJT | 1956 |
| 66 NW | Whitegate | BJT | 1953–56 |
| 66 NE | Bostock | BJT | 1956 |
| 67 SW | Hartford | BJT | 1956 |
| 67 SE | Northwich | BJT | 1956 |

# CHAPTER 1

# Introduction

## AREA AND INDUSTRIES

The district[1] described in this memoir, represented on Sheet 109 (Chester) of the 1:50 000 (England and Wales, New Series) Geological Survey map, forms a large part of central and west Cheshire; about 4 km² to the west of the River Dee, in the extreme south-west, fall within the Welsh county of Clwyd (Figure 1). In the west, the ancient city of Chester originated as the legionary headquarters of the Roman XX[th] Legion and evolved into the principal mediaeval market and ecclesiastical centre of north-west England. The old salt-mining town of Northwich is situated in the extreme north-east corner of the district, and Winsford, important in modern salt production, is now an important urban centre expanding on the lines of a designated new town. Notable rural areas include Delamere Forest and Vale Royal in the east, both formerly associated with an ancient royal hunting

1 Throughout this memoir the word 'district' refers to the area represented on the Chester (109) Sheet.

reserve, and the Eaton Hall estate in the west, the principal seat and centre of the Duchy of Westminster. The best known of the smaller, and essentially rural townships is Tarporley.

The situation of the district has ensured that many major roads, including four trunk routes, traverse it, and the surviving railway system includes the Hartford section of the electrified north-western main line as well as the Crewe-to-Holyhead line. An early phase in the development of industrial communications is represented by the Shropshire Union Canal, some 30 km of which still link the Chester-Wirral area with eastern England and the Midlands via Barbridge Junction in the south-east.

A large-scale chemical industry based on the local availability of Triassic salt has developed in the Northwich-Winsford area, though its supplies of brine now largely come from the Holford Brinefield in the Macclesfield (110) district to the east. The brine wells that once supported the salt trade at Winsford are abandoned, but salt mining is still important, and indeed the only active rock-salt mine in Britain is

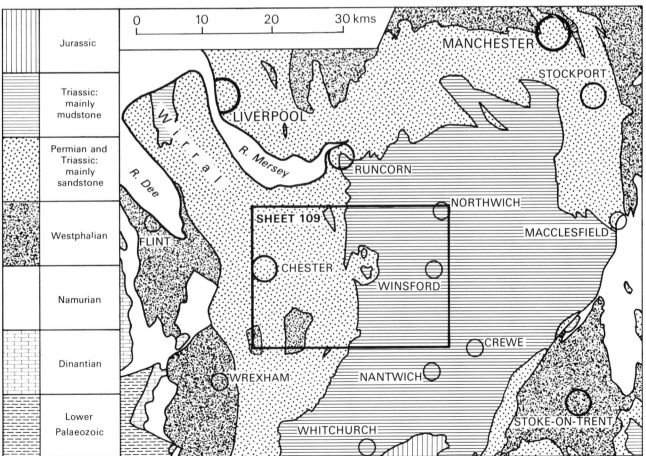

**Figure 1** Sketch map showing the location of the Chester-Winsford district and its relationship to the surrounding country

located north of the town. This mine supplies all the ground rock-salt used throughout the country to treat snow-and-ice-bound roads in winter. Chester has been, for many centuries, an agricultural market town, but in more recent decades, owing to its antiquities, riverside amenities, and zoological gardens, it has become a considerable recreational and tourist centre. The city itself has small brewing, engineering and other industries, and only a few miles to the north, on the northern boundary of the district, lies the great oil-refining complex of Thornton-le-Moors and Stanlow, as well as the large insulated electric cable plant at Helsby.

The nearness of the extensive, fertile, flat pasture land to the great conurbation of Merseyside has ensured that rural activity is dominated by dairy farming, but there is also a good deal of arable farming, especially on the more sandy soils. Forestry is a significant activity on the sandstone hills and spreads of glacial sands of the Delamere Forest area and of the Peckforton Hills. The Helsby Sandstone was once worked for building stone in large quarries along the outcrop from Helsby to Quarrybank near Tarporley and, on a smaller scale, in the Peckforton Hills, but quarrying has now ceased, as has the more ancient working of the Pebble Beds in and around Chester. Sand and a small amount of gravel are worked in large open pits in the area of Delamere Forest and Vale Royal, and to some extent in the areas south of Tarporley and south of Winsford.

Copious reserves of groundwater beneath the outcrops of the various members of the Sherwood Sandstone Group have been exploited, particularly by communities situated farther east on the mudstones and evaporites of the Mercia Mudstone Group. Northwich, Winsford, Middlewich and Crewe all used boreholes in the Delamere-Tarporley area. More distant towns, such as Warrington, exploited the Newton-Alvanley-Manley area; Stoke-on-Trent obtained a large supply from the Peckforton Hills and from nearby sites immediately to the south-west. Public supplies for rural areas in the west are drawn from the Ashton-Manley area and Chester supplements its surface supplies from a borehole at Plemstall. All these sources of supply are now controlled by the appropriate divisions of the regional water authorities. Very large supplies of groundwater for the oil refining industry are drawn from numerous boreholes, mainly in the Pebble Beds, west and south of Thornton-le-Moors.

The great aqueduct from Llyn Vyrnwy in North Wales to Liverpool has long supplied most places along its route from Peckforton in the south, through Tilstone Fearnall, Eaton, Cotebrook and Delamere to Kingsley in the north. The former Wallasey Branch from this aqueduct courses west from Hatchmere across the north-west part of the district. Apart from this aqueduct the largest surface source of water is the river Dee. Hurleston Reservoir on the south-east margin of the district, a major supply for this part of Cheshire, is fed via the Shropshire Union Canal from the Dee near Llangollen. Large supplies for the Wirral are drawn from the river at Heronbridge and Deeside, and for the Chester area at Barrelwell Hill. Very large supplies, mainly for Liverpool, are abstracted at Huntingdon and are carried by a major aqueduct across the north-western part of the district via Christleton, Barrow and Helsby to the Vyrnwy Aqueduct at Norton Tower. The river Weaver and the river Gowy are the only other major sources of surface water and large supplies for industry are taken from the Weaver at Northwich and from the Gowy below Thornton-le-Moors.

## PHYSICAL FEATURES

The most distinctive physical features of the district are the hills of Triassic sandstone that range from north to south across the middle of the district (Figure 2) and separate a low plain to the west from a slightly more elevated, but more dissected, plain to the east. The hilly belt is two to three miles wide and the more prominent tracts along it, with the levels above OD of their highest points, are the following: Alvanley Cliff – Birch Hill – Simmond's Hill, 155 m; Longley – The Yeld – Pale Heights – Eddisbury Hill, 175 m; Primrose Hill – Kelsborrow Castle – High Billinge, 179 m; Summerhouse Bank, 146 m; Beeston Castle, 154 m; Peckforton Hills, 198 m and Burwardsley Hill, 150 m. West of the hills the drift-covered plain is mainly underlain by the soft, mottled sandstones and Pebble Beds, its surface falling from around 40 m above OD down nearly to sea-level. The only hilly features within it are at Dunham-on-the-Hill, Little Barrow and Great Barrow. East of the hills lies a plain of thick glacial drift underlain by Triassic mudstones and evaporites, most of its surface lying between 80 m and 40 m above OD. There is lower-lying ground around Crowton in the north and along the river Weaver, which is incised 20 to 30 m below the general level of the plain.

The watershed between the Dee and Mersey river systems extends from the north-west corner of the district south-south-eastwards along gently elevated ground through Upton Heath, Christleton, Waverton and north of Tattenhall to the Peckforton Hills at Burwardsley. West of the watershed, small streams, the longest of which is Aldford Brook, flow to the river Dee. This river enters the district south-west of Churton, where it receives the river Alun from the west, and meanders northwards in a broad floodplain to Chester, where it swings abruptly westwards to flow through a shallow rock-cut gorge before reaching what was formerly the head of its estuary below the western walls of the city. Below Chester Weir the Dee is tidal.

A small, slow-flowing tributary of the Mersey, the river Gowy, and its side streams drain most of the rest of the ground west of the mid-Cheshire hills as well as a small area east of the Peckforton and Beeston hills. Below Plemstall, although only a relatively small stream less than 10 m wide, it flows in a broad floodplain, passing northwards out of the district near Thornton-le-Moors to join the river Mersey at Stanlow. It occupies an artificially straightened channel for most of its course from Huxley to Bridge Trafford and, north of Bridge Trafford, it is confined within artificial levees. A very small area immediately south of Helsby is drained directly to the Mersey by the stream known as Moor's Brook and Peckmill Brook.

Nearly all the ground east of the hills is drained northwards to the Mersey by the river Weaver and its tributaries. This river enters the district near Worleston as a stream 10 to 20 m in width; it meanders on a narrow incised floodplain northwards to its confluence with Ash Brook, and, between here and Winsford, its valley is extensively inundated as a

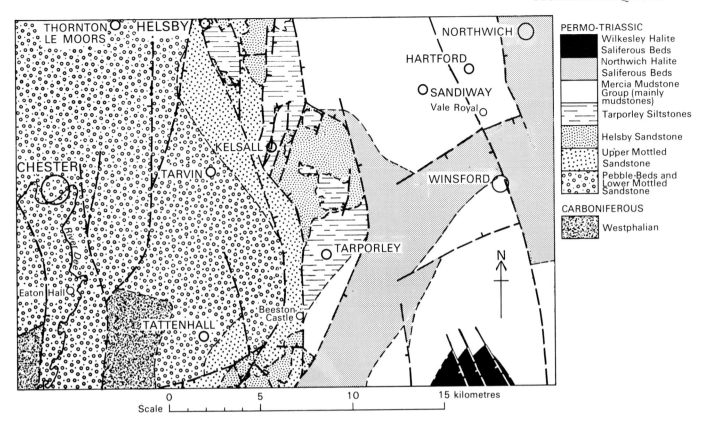

**Figure 2**  Sketch map showing an outline of the solid geology of the Chester–Winsford district

result of salt subsidence. The flooded areas are known as Winsford 'Flashes'. Below Winsford the river is canalised, and its reaches between Winsford and Northwich occupy a broad alluvial floored gorge, formerly a valley of considerable charm which acquired the ancient name of 'Vale Royal'. At Northwich the Weaver is joined from the east by the river Dane, a stream of comparable size meandering on a floodplain up to 300 m wide, the westernmost 4 to 5 km of which fall within the district. Between the Dane confluence and Weaverham, where it leaves the district, the river Weaver is a navigable river some 30 m in width.

The drainage of much of Delamere Forest in the Hatch Mere–Oak Mere–Petty Pool triangle was formerly into many self-enclosed hollows in the glacial sands. Most of these are peat floored and many have been artificially drained, but shallow areas of water, such as Hatch Mere and Oak Mere, are surviving lakes that lack effluent streams.

## GEOLOGICAL SEQUENCE

The sequence represented at outcrop is as shown on the inside front cover.

## CLASSIFICATION OF PERMIAN AND TRIASSIC STRATA

The classification of the above rocks employed in this memoir and on the new 1:50 000 sheet differs from that used previously. The earlier nomenclature had a long history.

Lower and upper subdivisions of the so-called 'New Red Sandstone Series' of rocks of eastern England and the east Midlands were correlated by Sedgwick (1835) with the lower, Buntsandstein (mottled sandstone), and the upper, Keuper (red marl) subdivisions of the German Triassic, because of the generally similar character and stratigraphical relationships of the two sequences. The terms 'Bunter' and 'Keuper' quickly gained acceptance in Britain and, by the mid-nineteenth century, these rocks in Cheshire were divided into a lower group of sandstones, usually called the Upper New Red Sandstone or 'Bunter', and an overlying group of gypsiferous and salt-bearing marls called the Upper Red Marls (Binney, 1841, p.37) and, a little later, the 'Keuper' (Ormerod, 1848).

As the early work of the Geological Survey in areas of Triassic outcrops progressed, further subdivision of the system was proposed by Hull (1860 and 1869), who divided the rocks of the central counties of England as follows:

Rhaetic or Penarth Beds
New Red Marl (Keuper Marl)
Lower Keuper Sandstone
  Waterstones
  Building Stones
  Basement Beds
Bunter Sandstone
  Upper Red and Mottled Sandstone
  Pebble Beds
  Lower Red and Mottled Sandstone

Hull considered that the 'Bunter' and 'Keuper' rocks were separated by a major disconformity and that the

Muschelkalk of the German sequence lay within a stratigraphic gap marked by this disconformity: he thus employed the terms 'Bunter' and 'Keuper' in an essentially chronostratigraphic fashion, drawing his disconformity within the sandstones that make up so much of the lower part of the sequence. The resurvey of Cheshire carried out by the Geological Survey in the 1950's and 1960's did little to upset this view, though it led to the identification of two saliferous formations within the 'Keuper Marl', which were termed the Lower Keuper Saliferous Beds and the Upper Keuper Saliferous Beds.

A major breakthrough came as a result of detailed palynological studies, first in the North Sea and then on land, in what had been previously thought to be essentially an unfossiliferous succession. As far as they effect the Cheshire district, these studies have established that there is no evidence for the regional break that Hull postulated. On the contrary it is clear not only that time-equivalents of the Bunter, Muschelkalk and Keuper of Germany are present in Cheshire, but that all occur within that part of the sequence formerly assigned to the 'Keuper' (though much of the Bunter is represented by unfossiliferous sandstones).

The results of the limited amount of palynological work carried out on Triassic deposits in this and neighbouring districts are summarised in Chapters 3 and 4. The ages assigned to Triassic formations on the basis of these results (Figure 11) should be regarded as approximate and provisional, because of the poor stratigraphic coverage afforded by the productive samples, and the sparse nature and poor preservation which renders many of the assemblages unsatisfactory for dating purposes.

Retention of the traditional British terminology would clearly cause much international confusion, and the present account makes use of much of the nomenclature put forward by Warrington and others (1980). This nomenclature is wholly lithostratigraphic in concept. Two major groups are proposed. The Sherwood Sandstone Group at the base includes the former 'Bunter Sandstone', together with that part of the 'Lower Keuper Sandstone' lying below the base of the 'Waterstones' of Cheshire, a horizon that in the Chester district is probably the best-defined datum-plane in the entire Permian and Triassic sequence. The overlying Mercia Mudstone Group includes all the overlying Triassic strata preserved within the Chester district, and is dominantly argillaceous or saliferous throughout.

It follows from this proposal that all the local terms that include 'Bunter' and 'Keuper' as prefixes should be abandoned. Consequently alternative names were proposed for each of these stratal divisions by Warrington and others (1980). Many of these we have accepted, though we have preferred informal terms to cover the former 'Lower Keuper Marl' and 'Middle Keuper Marl', since current research suggests that it may soon be possible to erect formal terms that will be based on variations in their lithology rather than their stratigraphic positions in relation to major salt formations. We have also accepted that the 'Waterstones' is inappropriate as a formational name. The 'Waterstones' of Cheshire are significantly different in lithology, palynological content and, presumably, age from the 'Waterstones' of the Midlands. The relationship of the two successions is still far from clear, and the continuation of a common terminology merely conceals the inadequate state of present knowledge.

We have, however, retained the subdivisions of the former Bunter as used on the 1965 edition of the Chester (109) Sheet, largely because, although inappropriate within the district, they are likely to remain in common useage for some time. The newer terminology has, however, been employed on the 1986 edition of the 1:50 000 Sheet with the older nomenclature given in brackets for ease of cross reference.

## GEOLOGICAL HISTORY

Significant direct evidence from within the district of its pre-Pleistocene history relates only to the Carboniferous Period and to parts of the Permian and Triassic periods. Ordovician rocks are known from only one borehole, and no post-Triassic solid strata are preserved.

Towards the end of Dinantian times within the Carboniferous Period, there was a major marine transgression into the district which started a long phase of sedimentation on a gently subsiding delta or sea floor that continued, with little interruption, throughout the rest of the Carboniferous. At first, mainly calcareous sediment was formed in the clean waters of a warm sea, but later, and throughout most of the Namurian and Westphalian epochs, terrigenous material, probably the deltaic detritus of rivers flowing from a north-easterly direction, was predominant. During late Namurian and most of Westphalian times the sedimentation tended to follow a rhythmic 'deep-to-shallow' pattern, the most characteristic feature of which was the regularly repeated formation of mud-flats that were colonised by dense forests of vascular cryptogams, the dominant plants of the time. Such phases produced the seatearths and coal seams of each sedimentary cycle. Towards the end of the Carboniferous Period, when the early Armorican earth movements were probably affecting both the erosive and depositional environments, most of the deltaic mud and silt brought into the district was laid down under oxidising conditions and gave rise to red sediments.

Deposition ended with the intensification of the Armorican orogeny, which disturbed the whole district, though it probably left the Carboniferous strata only relatively gently folded and faulted. The orogeny was followed by a long period of erosion, probably mainly in an oxidising environment, during the later phases of which the climate seems to have become semi-arid.

In late Permian and early Triassic times the district appears to have been part of a broad trough or depression, trending from south-south-east to north-north-west, occupied by the shifting distributory stream system of a continental drainage complex. Transported by both wind and water, sand was at first the major type of sediment to accumulate, but during early Triassic times there were repeated influxes of coarse pebbly sand, probably brought in from a southerly direction during periods of flood. This regime gave way gradually to another long period of mainly sand deposition.

The complementary processes of trough-bottom sagging and its infilling with coarse detritus continued, the depositional surface probably became almost flat and, towards the end of the early epoch of the Triassic Period, a widespread subsidence quickly converted the district into part of a gulf or lagoon, with restricted egress to or ingress from the sea, in which sediment comprised the muds and silts of a tidal estuarine environment. Phases of high salinity of the water became important factors in the depositional history. At first the gulf was fairly shallow and the salinity of the water fluctuated from more to less brackish as the balance of marine inflows, river water and evaporation varied. Then, in an environment of high evaporation, the salinity increased more permanently until much anhydrite was being precipitated along with red mud and silt; at times great areas of mud-flats seem to have suffered temporary desiccation. Finally the lagoon became so saline and locally sufficiently deep for thick-bedded evaporites, consisting mainly of rock-salt, to form. An early phase of rapid halite deposition was followed by a return to mainly anhydrite-impregnated terrigenous deposition, followed by another phase of rapid halite formation.

Evidence for the later Mesozoic history of the district is not preserved, but it probably remained part of a great subsiding basin of deposition until at least mid-Jurassic times, and may have been an area of deposition rather than of erosion until late in the Mesozoic era. The district was subsequently uplifted, tilted mainly towards the east, and locally warped into broad synclines. Following these earth movements a long period of erosion removed all the Mesozoic sediments, except those that now comprise the Sherwood Sandstone Group, from the western part of the district, but left higher Triassic sediments in the east. Erosion finally reduced the Triassic rocks to something like their present contours; the district became gently undulating terrain at around sea-level with a belt of more hilly ground running from north to south across the middle. The superficial deposits provide good evidence for there having been a Pleistocene phase (or phases) of cold climate and a subsequent amelioration to the temperate climate of today. Evidence of early Pleistocene glaciations is equivocal but in the later part of the Period the whole district was glaciated by a thick ice-sheet that moved southwards from south-west Scotland and across the northern part of the Irish Sea basin. There is widespread evidence of the melting and retreat of the ice-sheet, in the form of extensive deposits of sand and gravel which are themselves much overspread by boulder clay, presumably the final debris released from the decaying ice. The history of events since the ice disappeared some 12 000 years ago is mainly of the re-establishment of a thoroughly disrupted drainage, and is recorded by deposits of terrace gravel and alluvium along the stream courses. The steady amelioration of the climate during this period is recorded in the pollen content of the successive layers of peat preserved in the many waterlogged hollows. The final silting up of the landward end of the Dee estuary and probably also of the Gowy marshlands below Plemstall, has taken place since Roman times.

## SUMMARY OF PREVIOUS RESEARCH

Before the early systematic work of the Geological Survey, interest in the geology of this district focussed on three pro-

minent topics. These were: the salt deposits of the Northwich saltfield and their economic exploitation; the footprints, ripple marks, sun cracks and rain pittings that occur at various horizons in the Helsby Sandstone and the Tarporley Siltstones; and the marine shells that occur in drift gravels high above sea-level. Literature on salt in Cheshire dates back to the 17th Century, but the earliest noteworthy papers appeared towards the middle of the 19th Century and among these must be mentioned those of Ormerod (1848, 1869). Sir Philip Grey Egerton (1838, 1839) and other authors, described footprints from the Helsby Sandstone of this district and the same author, following Trimmer (1833), recorded marine shells in gravel at The Willingtons (1835) and at Norley Bank (1836).

The first systematic geological survey of the solid geology of the district was carried out on maps at the scale of one-inch-to-one-mile by E. Hull in the eighteen-fifties, and the results were published on one-inch-to-mile Old Series Sheet 80 SW (Chester, Delamere Forest) in 1855, and on Sheet 80 SE (Northwich, Middlewich) in 1858. Geological Survey Horizontal Section Sheet 43 was published in 1858; the eastern part of it runs from just north of Chester, east-north-eastwards to Pettypool Park, passing through the Pebble Beds ridge of Great Barrow and the 'Waterstones' outlier of Eddisbury Hill. Hull (1860) gave a general account of his 'new subdivisions' of the Triassic rocks of this and surrounding districts, a paper that led on to his more comprehensive memoir (1869) on the Triassic and Permian rocks of the midland counties of England.

During the period that elapsed between the work of Hull and its subsequent revision by Strahan and De Rance, who also made the first systematic survey of the drift geology of the district, many papers were published on the superficial deposits of this and adjoining districts. The more important of these were by De Rance (1870), Mackintosh (1873, 1880), Reade (1874) and Shone (1874, 1878a, b); Shone discovered foraminifera in the blue silt infilling of *Turritella* shells found in red boulder clay near Chester and up to this time, which was mainly before the work of Tiddeman in Lancashire, the glacial deposits of the Cheshire plain were believed by many to have been deposited in the sea. A short paper by Dawkins (1878) recorded the finding of a mammoth tooth in sands underlying thick boulder clay at Northwich. The Cheshire saltfield was the subject of a paper by Ward (1873) and, in the same year, Dickinson (1873) completed a report to the Home Secretary on land subsidence in salt mining areas, including Cheshire. The hydrogeology of the Triassic rocks of this, among other districts, was investigated by a committee of the British Association, under the chairmanship of De Rance, which published five reports on the results (1876–1879).

Around 1878 the solid geology of nearly all except the eastern quarter of this district was revised by Strahan and the drift geology was mapped in outline for the first time. New editions of one-inch-to-one-mile Old Series Sheet 80 SW (1882) were published, as well as a new edition of Horizontal Section Sheet 43 (1892). A memoir was written (Strahan, 1882), with an exhaustive bibliography of works from 1630 to 1881. A few years later De Rance revised the solid geology, and mapped in outline for the first time the drift geology of the easternmost quarter of the district. New editions of one-inch-to-one-mile Old Series Sheet 80 SE were published in 1890. De Rance continued his work in the saltfield with a description of an important borehole near Northwich (1895) and soon afterwards a sheet of Geological Survey Vertical Sections (Sheet 82, 1897) showing sequences of strata in the Cheshire saltfield was published.

During the first quarter of the present century there were further publications of varying scope on the Cheshire saltfield. Among these must be mentioned the voluminous work of Calvert (1915) who reviewed all aspects of the Cheshire salt industry. During the 1914–1918 war years Sherlock (1921) prepared a Geological Survey Special Report on rock-salt and brine.

A six-inch-to-one-mile survey of the Runcorn (97) Geological Survey sheet was carried out between 1935 and 1938; the work took in a narrow strip of country along the northern margin of the Chester district, and was briefly noted in the contemporary Summaries of Progress of the Geological Survey (Wright, 1936, 1937, 1938). Much new information was obtained, especially during the inter-war years, from boreholes drilled for water supplies into the Triassic sandstones, and two adjacent boreholes at the West Cheshire Water Board Pumping Station near Mouldsworth were described by Jones (1935). Information from wells and boreholes throughout the district was presented in summarised form by Wray and Earp (1944). In 1949 important conclusions about the deep structure of the district, based on regional gravity surveys by the Anglo-American Oil Company Ltd, were published by P. H. N. White.

In 1958–1959 several exploratory boreholes were drilled for the Geological Survey to supplement the surface mapping, particularly in areas of thick drift, and one of these proved the Milton Green inlier of Upper Coal Measures in the south-west. The results of this drilling programme were briefly noted in the contemporary Summaries of Progress of the Geological Survey (Stephens, 1958, 1959), and summarised logs are given in Appendix 1. The results of seismic traverses in this part of Cheshire were described by Mc-Quillin (1964).

Of three Geological Survey memoirs on nearby districts published between 1963 and 1968, that on the Stockport (98) Sheet (Taylor and others, 1963) included references to this district. The memoir on the Nantwich (122) Sheet (Poole and Whiteman, 1966) dealt with the geology of the southern margin of this district. It summarised new information about the Milton Green inlier of Upper Coal Measures in the south-west and gave details of the Geological Survey Wilkesley Borehole, which has proved to be so important in depicting outcrops of the Northwich and Wilkesley halites and associated mudstones on maps of this part of Cheshire. The memoir on the Macclesfield (110) Sheet (Evans and others, 1968) provided a more recent account of these halites and associated strata along the eastern margin of the district. In all three memoirs the sequence of glacial deposits, and of the events that they represent, were discussed at length.

The drilling of further wells and boreholes continues to supply new information about the sub-surface geology. A borehole was drilled in 1965 by the Esso Company Ltd to more than 1500 m into the Coal Measures inlier in the south-west and the National Coal Board has made eight deep exploratory holes in the western part of the district and in adjacent areas to the west and south. All these are referred to in Chapter 2. A sedimentological and stratigraphical study of the 'Keuper' Sandstone and associated strata of Cheshire was published by D. B. Thompson (1970a). As well as the general relevance of such a study to this district the author also dealt in some detail with the correlation of several sections, mainly cored borehole sections recorded by the Geological Survey through the Helsby Sandstone and the Upper Mottled Sandstone, in the Newton-Delamere-Beeston area.

JRE

# CHAPTER 2

# Carboniferous and older rocks

The Milton Green inlier of Upper Coal Measures was discovered in 1958 with the drilling of a Geological Survey borehole at Old Beachin Farm. The geology of the inlier and the background to its discovery was briefly described by Poole and Whiteman (1966, p.9). In 1965 the Esso Petroleum Company Ltd. further explored the inlier by means of the much deeper Milton Green No. 1 Borehole at Edgerley Farm, about 700 m WSW of Old Beachin Farm (Figure 3). This hole reached the base of the Carboniferous at a depth of about 1500 m and entered Ordovician sediments, which included a metre or so of highly altered igneous rock at about 1585 m. We are indebted to the Esso Petroleum Company for releasing these results for publication.

Between 1978 and 1980 the National Coal Board commissioned eight deep exploratory boreholes to prove Coal Measures beneath Permo-Triassic rocks to the east of the Flintshire-Denbighshire Coalfield. Four of these boreholes fall within the Chester district and four are only a short distance outside it to the west. At the time of writing, assessment of the results of these boreholes, in terms of geology and potential coal reserves, is still in progress, and we are grateful to the National Coal Board for permission to publish the limited stratigraphical summaries of them that are shown in Figure 3.

## ORDOVICIAN

At the Milton Green No. 1 Well (Figure 3) the base of the Carboniferous was fixed at about 1500 m on the basis of chipping samples and electrical logs. Two cores were taken, with incomplete recovery, between 1567.6 m and the bottom of the hole at 1588 m. Specimens from core No. 14 (1583.7 to 1588 m) are mostly dark, greenish grey or striped silty mudstone. In the lighter laminae of the striped beds the matrix and clastics have been partially replaced by dolomite; the darker laminae are composed of micaceous flakes (chlorite and illite) associated with abundant leucoxene. This association of leucoxene with chloritic minerals suggests a volcanic derivation, though no tuffaceous particles were observed. Between 1585 and 1586 m the core is a highly altered igneous rock which appears to be an albitised, chloritised and carbonated metabasite. Mr R. K. Harrison reported on specimen BLB 7241, from a depth of 1585 to 1585.4 m, as follows: 'Hard, pale grey and mainly of medium granularity, this consists of a subophitic mesh of slender (0.3 x 0.05 mm) to stout (0.7 x 0.4 mm) plagioclase laths with scattered microphenocrysts up to 1.5 mm across. Both phenocrystic and groundmass plagioclase laths are altered along cleavages to chlorite and contain dolomite and sulphide grains. The plagioclase is dominantly of sodic character with little apparent range of composition even in

the microphenocrysts. The maximum extinction on [010] is about 16° (average 10°) and refractive index = 1.539 ± .002; this indicates an approximate average composition near $Ab_{90}An_{10}$. All primary ferromagnesian minerals have been replaced by dolomite and chlorite, the latter exhibiting both anomalous blue and olive-brown polarisation colours. Leucoxene is abundant in the mesostasis. Subvesicular patches contain chlorite and carbonates. There is little accessory (or ? secondary) quartz. Core No. 13 (1567.6 to 1573.7 m) consists of grey or greenish grey silty and chloritic mudstone with calcareous veinlets and strongly-laminated bedding planes.

Dr S. G. Molyneux has recovered rare and poorly preserved acritarchs from core No. 13. Three samples (1569.1, 1570.6 and 1571 m) yielded specimens of *Acanthodiacrodium*, *Polygonium* and *Stelliferidium* which indicate a probable early Ordovician age. A sample from 1570.1 m contained a specimen of *Stelliferidium fimbrium?*, a species which ranges from the late Tremadoc to the Llanvirn and one from 1571.2 m yielded *Acanthodiacrodium*, *Stelliferidium* and *Coryphidium*. The specimen of *Coryphidium* is close to an undescribed species which occurs in the early Arenig graptolite Zone of *Didymograptus deflexus* of the Lake District. An early Ordovician age, possibly early Arenig, is suggested for the core between 1569.1 and 1571.9 m, but the strata below this remain undated.

Geochronological investigations by Mr F. J. Fitch and Dr J. A. Miller of two samples, igneous rock from core 14 and metamorphosed sediment from core 13, showed that they had similar geochronological histories, but provided no definite evidence of their age. It did show that they were folded and metamorphosed in a Silurian phase of the Caledonian orogeny between 439 and 414 million years ago, and that a strong Caledonian cleavage was imposed on the pelitic sediment around 414 ± 5 million years. It further showed that Variscan events caused weak partial overprinting of the rocks around 315 ± 8 million years ago.

## LOWER CARBONIFEROUS: DINANTIAN

At Milton Green No.1 Well (Figure 3) drilling fragments from the base of the Carboniferous at about 1500 m included some red calcareous sandstone, suggesting the possible presence of a local, very thin, red basement facies comparable with the Basement Beds of Creigiau Eglwyseg in the Wrexham (121) district (Wedd and others, 1927). The succeeding basal white limestones are overlain, at about 1469 m, by some 20 m of dolomitic limestones. About 1.22 m of these were cored to 1463.04 m, consisting of fine-grained dolomitic limestone, with productoid shells, *Koninckopora inflata* and a stunted and recrystallised foraminiferal fauna including the rare *Melatolla sp.*, overlain by coarse grey

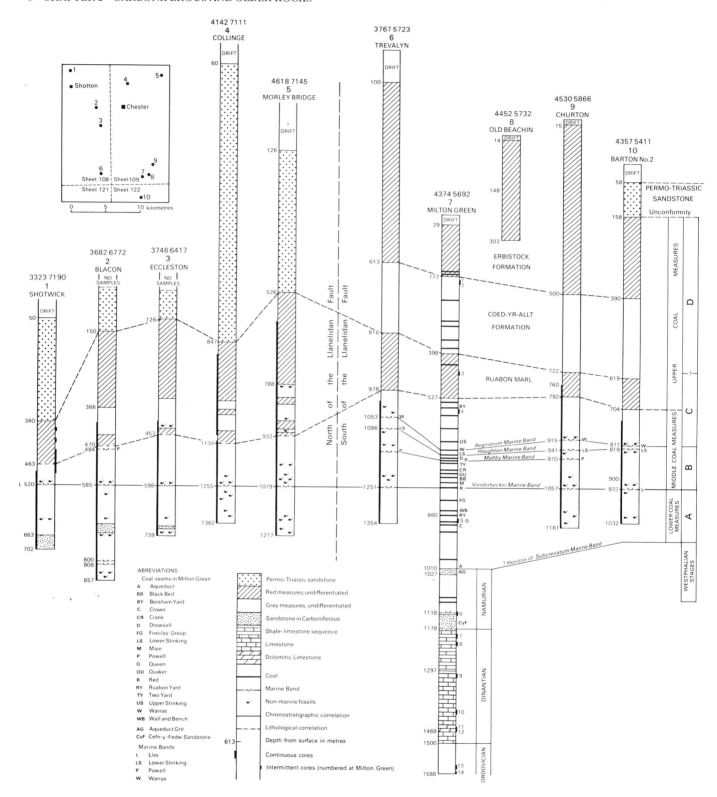

**Figure 3**   Comparative vertical sections of boreholes proving pre-Triassic rocks around Chester

limestone with dolomite in vugs, and productoid shells. An attempt to core the succeeding 6 m of strata between 1456 and 1462 m produced only worn pebbles of white and grey porous dolomite. The presence of bands of dolomite in this part of the Carboniferous Limestone is comparable with its widespread occurrence in North Wales, where the name Lower Grey and Brown Limestone is applied to the lithofacies.

The dolomitic limestones are succeeded at about 1450 m by grey and white limestones which extend up to about 1297 m, indicating a thickness of the order of 152 m for the main limestone unit of the Carboniferous Limestone. Core No. 10, taken from 1414.88 to 1418.23 m was pale grey-brown bioclastic and in places crinoidal limestone containing *Axophyllum vaughani*, *Linoprotonia sp.*, *Megachonetes sp.*, *Endostaffella sp.* and *Koninckopora inflata*. Core No. 9, taken from 1308.2 to 1317.04 m showed the following sequence.

|  | Thickness m |
|---|---|
| Limestone, grey to dark grey, coarse-grained, crinoidal; thin bituminous partings (some stylolitic); brecciated 25–30 cm below top | 0.76 |
| Limestone, pale grey-brown, coarse-grained; calcite veined; poorly preserved brachiopods near the base | 0.92 |
| Limestone, grey to dark grey, crinoidal; calcite veined | 1.22 |
| Limestone, greyish brown, crinoidal; numerous brachiopods; bituminous partings | 0.15 |
| Breccia of dark grey crinoidal limestone with some large ossicles, and pale brown very fine-grained limestone | 0.28 |
| Limestone, greyish brown, crinoidal (large ossicles); penecontemporaneous deformation and brecciation in places | 1.70 |
| Limestone, grey to pale greyish brown, rather fine-grained; penecontemporaneous deformation common | 3.81 |

Fossils from this core include *Lithostrotion portlocki*, *Syringopora* cf. *reticulata*, *Antiquatonia ?*, *Avonia sp.*, *Dielasma sp.* juv., *Fluctuaria sp.* juv., *Productus ?* (ss), *Schizophoria sp.*, *Spirifer bisulcatus* and cf. *Paladin*. A rich foraminiferal fauna includes *Loeblichia sp.*, *Nodosarchaediscus* (*Nodasperodiscus*) and *Archaediscus* at the *angulatus* stage.

It is reasonable to compare the limestones between 1297 and 1450 m with the lower part of the Upper Grey Limestone and the White Limestone which, at Minera, underlie the so-called Main Shale (Earp, 1958, p. 46). If this comparison is valid, the thickness of the measures here is much reduced, either by faulting or by lateral attenuation.

Between 1297 m and 1280 m these limestones are overlain by dark grey silty and calcareous shales with a calcareous sandy bed at the top. These are succeeded by a sequence of limestones, argillaceous limestones, shales and thin silty or sandy beds that extends up to about 1178 m. Core No. 8 was taken between 1218.29 m and 1222.25 m, giving the following sequence.

|  | Thickness m |
|---|---|
| Limestone, dark grey to black, with *Dielasma sp.* and *Gigantoproductus sp. giganteus* group; ?sandy at base | 0.83 |
| Mudstone, black calcareous; abundant shelly fauna including *Antiquatonia sp.*, *Rugosochonetes sp.*, and *Spirifer sp.* | 0.38 |
| Limestone, black; fine-grained matrix; brachiopods including *Antiquatonia sp.* and *Spirifer sp.* | 0.08 |
| Mudstone, black calcareous; abundant shelly fauna including *Spirifer sp.* and *Leiopteria sp.* | 0.05 |
| Limestone, black hard; fine-grained matrix; irregular base; brachiopods common including *Antiquatonia?*, a smooth spiriferoid and *Spirifer sp.* | 0.36 |
| Mudstone, black calcareous; brachiopods common, including *Antiquatonia sp.*, *Brachythyris sp.*, *Eomarginifera sp.*, *Lingula sp.* and an orthotetoid | 1.95 |
| Limestone, black with crinoid debris, *Productus sp.* (ss), *Spirifer bisulcatus* and fish debris | 0.28 |
| Mudstone, black calcareous with irregular limestone pockets; brachiopods including *Spirifer bisulcatus* | 0.33 |

Higher in the shale-limestone sequence core No. 7, taken between 1194.21 and 1203.05 m, consisted of 4.27 m of fine-grained dark grey limestone with muddy layers, overlain by 4.57 m of very fine-grained dark grey limestone, becoming less fine-grained upwards. The core yielded *Antiquatonia sp.*, *Avonia ?*, *Dielasma sp.*, *Eomarginifera ?*, *Productus sp.* (ss), *Rugosochonetes sp.*, *Spirifer sp.*, *Paraparchites sp.* and a fish spine.

It is likely that the beds between 1297 m and 1178 m represent the upper part of the Upper Grey Limestone and the Sandy Limestone of the North Wales outcrop. If so, the thicknesses in the borehole approximate to the minimal thicknesses of these beds at outcrop.

Mr M. Mitchell and Dr N. J. Riley report that the faunas in the Dinantian of the Milton Green Borehole are mainly of wide ranging Viséan species. The fauna from core No. 10 with the association of *Endostaffella sp.* and *Koninckopora inflata* indicates an age range from mid-Holkerian to basal Brigantian; the foraminifera in core No. 9 are of Brigantian age and *Gigantoproductus sp. giganteus* group in core No. 8 indicates an early Brigantian age.

The tentative lithological and faunal correlations given above suggest, by comparison with George and others (1976, p. 33), that these beds are mainly of Asbian and early Brigantian age.

## UPPER CARBONIFEROUS: NAMURIAN

Between 1139 and 1178 m drilling samples from Milton Green No. 1 indicated the presence of medium-grained light grey sandstones with silty and shaly layers. The position of this essentially sandstone formation in the sequence suggests that it may represent the Cefn-y-fedw Sandstone of the Millstone Grit Series of North Wales, but its thickness of about 49 m is less than a quarter of that sandstone at outcrop.

Between 1010 and 1134 m drilling samples indicate a sequence of shales, siltstones and sandstones, at the top of which a bed of medium- to coarse-grained light grey siliceous sandstone, about 17 m thick, is overlain by a thin coal seam. Strata near the base of this sequence, cored between 1134.47 and 1136.6 m (core No. 6) are as follows, and include a marine fauna not diagnostic of any specific marine band.

|  | Thickness m |
|---|---|
| Sandstone, very fine-grained, white; passing down into grey micaceous siltstone with shaly beds; penecontemporaneous faulting; plant fragments at base | 0.76 |
| Mudstone, dark grey; scattered pyritic concretions; abundantly listric in lower half, with fish remains, very abundant near base; upper half with *Serpuloides stubblefieldi*, *Lingula sp.* and fish remains including *Elonichthys ?*, *Pleuroplax* tooth, *Rhabdoderma sp.*, *Rhadinichthys sp.* and *Rhizodopsis sauroides* | 0.61 |
| Siltstone, grey; ? bioturbation and pyritic concretions near top | 0.30 |
| Mudstone, dark grey; *Orbiculoidea nitida*, and fish remains | 0.05 |
| Siltstone, striped; slumping and pyritic concretions | 0.10 |
| Mudstone, dark grey; siltstone bands towards base; *Orbiculoidea nitida* and fish remains in top 2 cm | 0.30 |

The strata between 1027 and 1134 m are correlated with the Holywell Shales by the Esso Company, and the overlying sandstone and coal seam respectively with the Aqueduct Grit and the Aqueduct Coal of the Denbighshire Coalfield (Wedd and others, 1928). The Subcrenatum Marine Band, internationally accepted as the base of the Westphalian division of the Carboniferous, lies close above the Aqueduct Grit in Denbighshire (Wood, 1937, p.14), and a coal commonly intervenes between the sandstone and the marine band. The stratal boundary between Namurian and Westphalian is, therefore, tentatively placed at the roof of the coal named as Aqueduct in Figure 3.

## UPPER CARBONIFEROUS: WESTPHALIAN (COAL MEASURES)

In Westphalian times the Chester area lay within an extensive depositional basin in which sedimentation kept pace with subsidence, creating a shallow-water estuarine environment which was periodically invaded by saline waters from the open sea. The basin was bounded to the south by the Wales-Brabant Island or Midland Barrier (Trueman, 1947, p. xxiv; Wills, 1956, pp.10–11) and covered the whole of present-day northern England. It is referred to as the Pen-

nine Province of Westphalian deposition by Calver (1968, pp.2–3). The nearest extensive outcrops of the rocks of this period are in the Denbighshire Coalfield to the west and a useful summary account of the Westphalian stratigraphy there is given by Calver and Smith (1974, pp.169–183). The Milton Green inlier has been referred to above; within it the Erbistock Formation, high in the Westphalian, lies directly beneath drift.

It has long been considered that the Westphalian rocks of north-east Wales are continuous at depth, beneath the Permian and Triassic rocks of Cheshire, with those of Lancashire, and of north-east Cheshire and north Staffordshire on the other side of the structural basin. The recent exploratory boreholes confirm that Westphalian strata continue eastwards beneath the unconformable Permo-Triassic strata, though the structural relationships are not simple.

White (1949) noted, about five miles south of Chester, an east–west zone of steep gravity gradients aligned with the Llanelidan Fault (p. 57 and Figure 20). This fault is the easterly continuation of the ancient Bala-Bryneglwys fracture, which abruptly terminates the Wrexham Coalfield at its northern end. The proved sections in the concealed coalfield are shown in Figure 3. There are significant differences between the boreholes north of the Llanelidan Fault and those south of the fault where the Trevalyn (6), Milton Green (7), Churton (8) and Barton No.2 (10) boreholes prove most of Westphalian A, all of B and C, and the lower part at least of D, comprising Productive Coal Measures, Ruabon Marl, Coed-yr-Allt Formation and Erbistock Formation. The thickest continuous Westphalian sequence, 1254 m, was proved by the Trevalyn Borehole, which entered Erbistock Formation at 100 m below surface and stopped drilling at 1354 m, some 50 m short of the base of Westphalian A. This thickness and sequence are similar to those in the exposed Denbighshire Coalfield south of Wrexham (and south of the Llanelidan Fault), where some 900 m of red measures rest upon 500 m of grey measures (Calver and Smith, 1974, figures 42, 43).

Boreholes in the concealed coalfield north of the Llanelidan Fault found a much more restricted Westphalian sequence beneath the unconformable Permo-Triassic cover. At the Shotwick Borehole (1) the Lower Mottled Sandstone lies a mere 180 m above the Vanderbeckei (Llay) Marine Band; at the Eccleston Borehole (3) the interval is 472 m; to the north-east at the Collinge (4) and Morley Bridge (5) boreholes it is 408 and 553 m, respectively. In all cases in this northern group of boreholes there is a zone of 'red beds' beneath the base of the Lower Mottled Sandstone; this zone varies in thickness between 120 m (Shotwick) and 406 m (Morley Bridge).

One explanation of these facts is that the full Westphalian sequence once formed a thick continuous sheet over the area, then strong post-Carboniferous block-faulting followed by vigorous erosion caused up to 1000 m of strata to be removed north of the Llanelidan Fault, at the same time preserving almost the full sequence to the south. The zone of 'red beds' beneath the Permo-Trias in the northern group of boreholes might then be seen, at least in part, as *in situ* reddening of originally grey Coal Measures beneath an arid terrestrial erosion surface over which the Lower Mottled Sandstone was eventually deposited: the correlation shows, however, that the red measures probably locally include the Ruabon Marl though the distinction between primary red beds and secondarily-reddened primary grey beds is generally impractical to make.

An alternative explanation is that there was movement along the Bala-Llanelidan Fault at intervals during the Westphalian, particularly during late in the epoch. To the north of the fault, deposition of the higher beds was interrupted, and there may have been occasional emergence and erosion of existing strata. The presence of the 'Symon Fault' unconformity in the Coalbrookdale Coalfield some 50 km to the south-east indicates such intra-Westphalian activity along the south-western fringe of the Pennine Province of Coal Measures deposition. On balance, this view seems unlikely because there is no sign of any systematic stratigraphical thinning in the remaining grey Westphalian beds in sequences north of the Llanelidan Fault or of any sudden thickening across the fault. For instance, the vertical interval between the Vanderbeckei and Maltby (Powell) marine bands at the Blacon Borehole north of the fault is somewhat greater than the same interval in the Trevalyn and Churton boreholes (Figure 3) south of it. The fault is thus considered to be an ancient fracture that moved again in late-Westphalian or early-Permian times.

Views on the relationships between the red and grey measures in the North Wales coalfields were summarised by Calver and Smith (1974, p.179). They point out the similarity of the succession of Ruabon Marl, Coed-yr-Allt Formation and Erbistock Formation in North Wales to the Etruria Marl, Newcastle Formation and Keele Formation sequence in North Staffordshire, where research by BGS has shown that the red beds are a facies, the onset of which occurs at different times in different places. Thus, within North Staffordshire the lower part of the red Etruria Marl passes laterally into the grey Black Band 'Group', and part of the red Keele Formation in some areas is the lateral equivalent of the grey Newcastle Formation elsewhere (Evans, 1966, p.54). Likewise, the Ruabon Marl of the Denbighshire Coalfield south of Llay is thought to be a possible stratigraphical equivalent of the grey Buckley Fireclay Group of south Flintshire. These views apply equally well to the red measures proved in the five boreholes south of the Llanelidan Fault. In the absence of firm faunal evidence the base of the Upper Coal Measures in these five boreholes is put for convenience at the base of the Ruabon Marl. North of the Llanelidan Fault in the ground proved by bores shown in Figure 3, sections 1 to 5, relationships between red and grey measures are complicated by the presence of a zone of oxidation which is though to have developed beneath an arid land surface over which the Permo-Triassic rocks were later deposited. In this zone both the free carbon and ferrous iron compounds, which together are responsible for the overall grey or greyish buff colouration in the grey measures, have been oxidised, the carbon being totally removed and the ferrous iron converted into red oxide. Original or primary red beds are virtually unaltered within the zone, while the grey beds assume an appearance which may cause them to be taken for primary red beds until the problem is appreciated, particularly where the oxidation zone is thick. The phenomenon was noted by Bailey (1926), its explanation was developed by Trotter (1953) and applied by Taylor (1961) to the Cumberland Coalfield where problems relating to red beds are similar to those in North Wales. In North Wales most studies of the red beds – grey beds relationships took place before the seminal work of Bailey and Trotter, with the result that, as in Cumberland and elsewhere, unconformities implying intra-Carboniferous earth movements have been invoked to explain phenomena that may be due to reddening beneath a post-Carboniferous landscape.

## Details

It is convenient to present these stratigraphical details under the same main headings as were used by Calver and Smith (1974), namely, 'grey measures' and 'red measures'.

### GREY MEASURES

South of the easterly extension of the Llanelidan Fault the Westphalian grey measures proved in boreholes equate with the Lower and Middle Coal Measures, so far as can be determined. They include Westphalian A, B and the lower part of C, the boundary between the Upper Similis-Pulchra and Phillipsii Chronozone being arbitrarily taken at the base of the Ruabon Marl in the absence of an equivalent of the Top Marine Band. (For an explanation of the classification of the Coal Measures of the region see Calver, 1968, plate 1). These strata are sometimes referred to as the Productive Measures because they contain the workable coal seams. They consist of a repeated sequence of cyclic units (cyclothems), each of which ideally includes a marine shale bed at the base, followed by non-marine shale or mudstone, sandstone, a rootlet bed or seatearth and then a coal seam. In most cyclothems one or more of these elements may be missing.

The Milton Green section shows many coal horizons which have been correlated with named seams in the North Wales Coalfield, (Figure 3), though in view of the nature of the borehole record these identifications must be regarded as tentative. The sequence has been correlated with the National Coal Board's exploratory boreholes on the basis of Westphalian stages (identified by the relevant marine bands) in the Lower and Middle Coal Measures, and on the basis of the lithological sequence in the Upper Coal Measures, where the holes are largely uncored and chronological markers are rare. Above 1010 m, drilling samples (Appendix 1, p.103 and Figure 3) indicate a thickness of about 484 m of mainly grey shales, siltstones and sandstones with many coals. Core No. 5, taken between 873.6 and 878.1 m, was of grey to dark grey siltstone and silty shale, some of it slickensided, and included a thin coal, equated with the Queen seam of the exposed coalfield.

A varied but mainly non-diagnostic assemblage of spores was obtained from the core at 876.3 m by Dr B. Owens. The most significant among these are *Dictyotriletes bireticulatus* and *Radiizonates (Densosporites) striatus* which suggest a late Communis to early Modiolaris Chronozone age. Much higher in the Coal Measures, core No. 4 between 559.9 and 564.5 m, was of grey to dark grey claystone and shale with bands of coaly shale, the lower and middle parts of the core being brecciated and slickensided.

General inter-borehole correlations can be made between the three N.C.B. boreholes south of the Llanelidan Fault and the named seams in the Milton Green sequence, using accepted correlations in the Denbighshire Coalfield. There is difficulty in fixing the base of the Lower Coal Measures (Westphalian A) at Milton Green because the Subcrenatum Marine Band is not recorded (see above). The N.C.B. holes (Figure 3) went at most only 106 m below the Vanderbeckei Marine Band and may not have bottomed the Lower Coal Measures, which in the Denbighshire Coalfield are around 180 m thick.

The identification by the N.C.B. in their boreholes at Trevalyn, Churton and Barton No. 2 of the Vanderbeckei and Aegiranum (Warras) marine bands enable precise bases to be drawn for Westphalian B and C respectively in these holes. There is no direct evidence to help in fixing the base of the 'Upper Coal Measures', but indirect faunal evidence from the Flintshire and Denbighshire coalfields suggests that the horizon of the Cambriense Marine Band lies close to the base of the Ruabon Marl (see Calver and Smith, 1974, pp.177-178).

North of the Llanelidan Fault a very restricted sequence of grey Productive Measures is proved in the Shotwick, Blacon, Eccleston, Collinge and Morley Bridge boreholes of the N.C.B. (Figure 3), beneath the transgressive base of red or reddened Westphalian strata. It seems possible that the lower of two *Lingula* bands at 800 and 808 m respectively in the Blacon Borehole may represent the Subcrenatum Marine Band marking the base of the Lower Coal Measures, although Calver (1968, figure 6) shows this bed in the Flintshire and Denbighshire coalfields to be in a goniatite-pectinoid facies. The Llay Marine Band is at 585 m, so this interpretation gives a thickness of 223 m for the Lower Coal Measures (Westphalian A) at Blacon. The Eccleston, Collinge and Morley Bridge boreholes probably stopped short of the Westphalian base. The Shotwick hole proved a sandstone not dissimilar in lithology and in its distance below the Vanderbeckei Marine Band to one named 'Aqueduct Grit' by the Esso Company in the Milton Green Borehole, but it is believed that this borehole stopped short of the Westphalian base.

Above the Vanderbeckei Marine Band the thickness of measures unaffected by reddening diminishes westwards from 147 m at the Morley Bridge Borehole to 59 m at Shotwick. The Maltby Marine Band was identified at 484 m in the Blacon Borehole, 5 m below the base of the reddened zone.

RED MEASURES (INCLUDING COED-YR-ALLT FORMATION)

South of the Llanelidan Fault the three main elements of the Upper Coal Measures are present, comprising Ruabon Marl, Coed-yr-Allt Formation and Erbistock Formation.

In the Denbighshire Coalfield the Ruabon Marl consists of red or purple mudstone with green or yellow banding and mottling. Thin beds of grey or black mudstone and coal smuts are occasionally found. At Milton Green (Figure 3) drilling samples between about 527 and 398 m indicate that the grey coal-bearing measures are succeeded by a group of mainly brown, red, purple or variegated strata in which coaly horizons are rare. Core No. 3, taken between 451 and 455 m was of grey-green slightly pyritic shale and siltstone with a thin seam of coal. It is a reasonable presumption that these strata represent the Ruabon Marl of the Denbighshire Coalfield. The lowest 32 m of the Ruabon Marl was cored at the Churton Borehole (Figure 3). The main lithology is recorded as red-brown mudstone with grey mottlings, with thin layers of light grey mudstone, silty mudstone, siltstone and sandstone.

The thickness attributed to the Ruabon Marl, chiefly on the basis of chipping samples, is 162 m at Trevalyn, 129 m at Milton Green, 85 m at Barton No. 2 and 70 m at Churton. This indicates a progressive easterly thinning from the 200 m recorded around Wrexham (Calver and Smith, 1974, p.178).

The Coed-yr-Allt Formation in the exposed Denbighshire Coalfield consists chiefly of grey mudstones, sandy mudstones and sandstones. Parts of the succession have red and purple bands and mottling. There is a similar cyclic sequence to that in the grey productive measures, but coals are thin and there are limestone bands, one of which forms the basal member of the formation in places. Non-marine bivalves are recorded, including *Anthraconauta phillipsii* and *A.* aff. *tenuis*. The base of the formation approximates to the Westphalian C-D boundary.

At Milton Green, a thick unit of mainly grey strata with coaly

horizons, between about 398 and 177 m, is correlated by the Esso Company with the Coed-yr-Allt Formation of North Wales. As well as thin coal seams, these beds include several thick sandstones and silty sandstones, especially from 396 m, 295 to 275 m, and 213 to 173 m. Cores Nos. 1 and 2, between 192 and 198 m were mainly of fine-grained grey-green micaceous sandstone and silty sandstone with some red sandy siltstone at the top. Similar beds of comparable thickness were identified on the basis of chipping samples in the three N.C.B. boreholes. Recorded thicknesses in all four deep holes were: Trevalyn 203 m, Milton Green 221 m, Churton 222 m and Barton No. 2 229 m, this shows a progressive eastward thickening from the 90 to 160 m recorded in the Denbighshire Coalfield to the west, a trend diametrically opposite to that noted in the Ruabon Marl.

The Erbistock Formation is present in the Trevalyn, Milton Green, Churton, Old Beachin and Barton No. 2 boreholes (Figure 3), the thickest sequence being the 513 m proved at Trevalyn. All the sections are presumed to be incomplete either beneath drift or the unconformable base of the Permo-Trias, but there is as yet no way of estimating the extent of the erosion. The formation attains a thickness of 900 m in the Denbighshire Coalfield (Calver and Smith, 1974, p.177).

The highest measures at Milton Green, between the base of the drift and about 177 m depth, are recorded as mainly brown, grey or greenish grey claystone and siltstone, and are correlated by the Esso Company with the Erbistock Formation although no cores were taken.

The Old Beachin Farm Borehole (Appendix 1, pp.105–106 and Figure 4) obtained cores of red, purple, green and grey Upper Coal Measures to a total depth of 302.74 m. Below 294.08 m are purplish or greenish grey mudstones and siltstones, and 106 cm of silty sandstone with a sharp base at 302 m. These are overlain by 14.53 m of mauve medium- to coarse-grained sandstone, the sharply defined base of which is coarse clay-gall conglomerate. At the top of this sandstone, at 279.55 m, there are 38 cm of dark grey carbonaceous shale that yielded much organic debris, including the crustaceans, *Falsirostria (Leaia) bristolensis* and *Euestheria simoni*.

Between this important horizon and a depth of about 148 m the strata display some elements of a cyclic pattern in which units of purplish mauve mudstone, or mauve and greenish grey siltstone are separated by a metre or two of seatearth overlain by greyish or greenish plant-bearing or carbonaceous shale. Such seatearth–carbonaceous shale horizons, analogous to the seatearth–coal elements of a normal grey Coal Measures cycle, are present at about 266 m, 252 m, 247 m, 243 m, 238 m, 219 m, 200 m, 181 m, and 159 m. Ascending this sequence the proportion of grey or greenish grey material tends to diminish, and above about 148 m the strata are red, purple or variegated mudstones with greenish stripes and 'fish-eyes', subordinate silty sandstones and a few thin beds of sandstone. The mudstones include a few pebbly horizons and autobrecciated layers. Plant fragments occur at several levels in the mudstones and, between 64.6 and 66.2 m, red mudstones yielded *Spirorbis sp. Euestheria simoni*, ostracods and plants.

Before the drilling of the nearby Milton Green No. 1 there was little evidence for a more precise correlation of the Old Beachin section than to equate the measures with some part of the Erbistock Formation. However, as the Milton Green Borehole apparently entered Coed-yr-Allt Formation at about 177 m, and the strata at Old Beachin include more greyish and greenish material towards the base of the borehole, it seems likely that they lie low rather than high in the Erbistock Formation.

North of the Llanelidan Fault the base of the zone of reddening is sharply transgressive, its distance above the Vanderbeckei Marine Band diminishing westwards from 147 m at the Morley Bridge Borehole to 57 m at Shotwick. The base of the zone is not clear-cut, and in the lower part there is alternation of grey and red measures similar to that in the 'zone of partial oxidation' described by Trotter

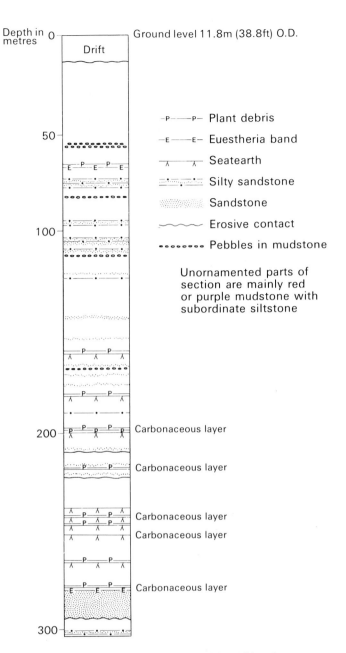

**Figure 4**  Generalised vertical section of the Erbistock
Formation in the Old Beachin Farm Borehole

(1953, p. 11 and plate 1), in south Lancashire where, as in these
boreholes, the alternating measures lie between unaltered grey beds
below and completely oxidised strata above, and are strongly tran-
sressive, in that case from Upper Coal Measures down to
Namurian. Sections 1 to 5 in Figure 3 summarize the facts as they
affect this district. It is apparent that much of the red strata occupies
the position of the Ruabon Marl. How far this facies transgresses
into the Middle Coal Measures is as yet uncertain.            BJT

# CHAPTER 3

# Permian and Triassic: Sherwood Sandstone Group

The sediments which now form the lower part of the Permo-Triassic succession in the Shropshire-Cheshire-Lancashire lowlands are believed to have accumulated in a broad continental trough or depression trending approximately SSE–NNW. The western flank of this depression probably lay along the eastern foothills of a rugged mass of mountains, the degraded remnants of which now form North Wales. Its eastern flank was less clearly defined by a range of low hills that lay approximately along what is now the Pennine axis and that separated the Cheshire part of the trough from the margin of a great, gently subsiding plain that lay away to the east and for which the name North Sea Basin is used.

The 'Cheshire' trough, which was not less than 30 km in width, seems to have formed the lower reach of a continental drainage complex, for it was occupied almost continuously during the accumulative period by a system of sediment-laden watercourses and wadis that were probably flowing from a mainly southerly direction. These streams brought in large quantities of detritus, much of which may have been derived from the frontal ranges of the newly formed Armorican mountains of southern Britain and northern France, into which many deep gorges were presumably cut in late- and post-orogenic times; they discharged the detritus onto a wide and constantly changing floodplain. Periodically the flow of water was strong enough to bring in coarse pebbly sand; at the other extreme, the area occasionally became so arid that winds shifted, eroded and redeposited the sand.

The arenaceous unit that results from this fluvial and aeolian activity comprises the Sherwood Sandstone Group (see p. 4). Its deposition began during Permian times and ended within the Triassic when the nearly flat floodplain was converted into a lagoon, the waters of which received generally fine-grained sediment and became increasingly charged with sulphates and chlorides.

The sequence has long been subdivided into four, and this subdivision has been retained in this account (Figure 5). In upward order the formations are the Lower Mottled Sandstone, the Pebble Beds, the Upper Mottled Sandstone and the Helsby Sandstone, this last name replacing the earlier 'Keuper Sandstone' (see p. 4) and taking its name from Helsby, where the strata are excellently exposed.

Precise definitions of these formations are difficult, if not impossible, to produce. In concept the Pebble Beds and the Helsby Sandstones are generally pebbly units, while the others are non-pebbly. In practice, pebbles are not confined to the pebbly units, while these themselves contain substantial non-pebbly sections. Arguably the units are properly regarded as facies rather than formations, but there are regional advantages in taking the traditional view. Inevitably, however, the limits taken for the formations are largely arbitrary. In general the lowest and highest levels where pebbles are common are taken as the boundaries of the Pebble Beds, and the incoming of another pebbly-unit as

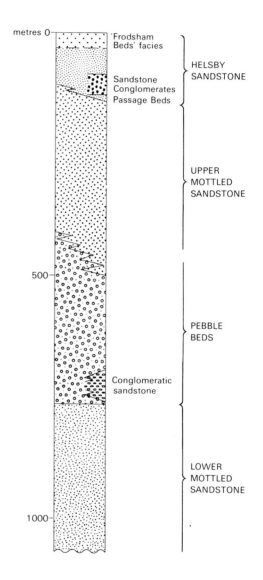

**Figure 5**   Generalised vertical section of the Sherwood Sandstone Group

the base of the Helsby Sandstone. Nevertheless, where several good cored sections are available in the same area, major compromises have to be made to maintain the formational concept. It is also practically certain that the formations are diachronous, and for this reason, as well as the arbitrary nature of their definitions, the significance of such thickness variations as have been established is far from clear.

The arbitrary definitions of the formations also have implications in the structural interpretation of the district. In particular, the magnitude and even the positions of some of the postulated N–S faulting within the Group depend on the internal consistency of the formational boundaries.

## LITHOFACIES AND SEDIMENTATION

No sedimentological studies were carried out during the primary survey of the Chester district, which was completed in 1958. Since then there has been increased interest in such research and significant studies on Triassic sediments have been undertaken by several workers. The results most relevant to the Chester district are those of D.B. Thompson (e.g. 1970a, b). He has studied the Upper Mottled Sandstone and the Helsby Sandstone in considerable detail (Thompson's basal member of the Helsby Sandstone—the Thurstaston Member—is considered in this account to fall within the Upper Mottled Sandstone), though his interpretations are also applicable to the other formations in the Group, since the strata comprise alternations of a limited number of fluvial floodplain and aeolian facies.

The Lower Mottled Sandstone is not sufficiently well exposed within the district for sedimentary studies to be worthwhile. In particular it is uncertain whether there is a recognisable unit within it that can be equated directly with the Collyhurst Sandstone of South Lancashire.

The Pebble Beds include some sequences that show a rather poorly developed 'fining-upwards' cyclicity, in which a metre or two of abundantly pebbly sandstone is followed by finer-grained sandstone with only scattered pebbles or none at all; this may be followed by another sandstone with a coarse pebbly base. Locally a bed of finer-grained sandstone passes up into a thin layer of sandy or silty mudstone. In many sections, however, there is little evidence of a cyclic pattern, and scattered pebbles may occur throughout beds of cross-bedded sandstone up to 10 or even 15 m thick. Such large elements may be interbedded with almost equally thick beds of similar but pebble-free sandstone. Beds that could be described as 'shingle-gravel', which are so abundant in the Midlands, are relatively few, seldom thick, and generally lenticular. The 'flood-gravel' regime of the streams that produced the soft shingly and cyclic Pebble Beds of the Midlands was clearly considerably less energetic in Cheshire, as shown by the widespread dispersal and chance entombment of pebbles in the sand.

Thompson finds that the bulk of the Upper Mottled Sandstone consists of two main lithofacies, the major predominating over the minor in a proportion of roughly 3 to 2. The major constituent is 'foxy red-brown, generally fine-, occasionally medium- or coarse-grained, trough, lenticular rarely tabular, planar cross-bedded, often deformed soft sandstone. Sets are usually between 0.5 and 2 m thick. Mica may be common or rare, in which latter case millet seed grains are common. Argillaceous matter may be found'. He infers that the part of this facies which is rich in mica and large galls was undoubtedly deposited in a low-sinuosity fluvial environment, but that some of the well sorted beds with millet-seed grains may be aeolian. The minor constituent of the Upper Mottled Sandstone is described as 'darker interbedded flat- and cross-bedded argillaceous soft sandstone. Mudstone laminae may be pre-

sent. Rarely, the flat beds have ripple or wavy bedding, though most appear structureless'. Thompson provisionally infers that the facies may be top-strata laid down at the margins of braided river channels. A third, but very minor constituent of the Upper Mottled Sandstone comprises thin layers of so-called 'marl'. These consist of 'chocolate-brown interbedded shale, siltstone and fine-grained sandstone. Occasional micaceous horizons, millet-seed lenses, injection structures, mud-cracks and blocky bedding are present. The facies is usually very thin but can reach 14 m. It is markedly impersistent laterally'. It is inferred that these strata were deposited in temporary pools between dunes and within or outside river channels while the more considerable thicknesses formed in flood-basins.

As the main constituents of the Helsby Sandstone (his 'Keuper Sandstone Formation'), Thompson recognises an underlying 'Delamere Member' and an overlying 'Frodsham Member'. The bulk of the 'Delamere Member' consists of 'red, secondarily buff or white, lenticular planar or trough cross-bedded, mostly coarse, ill-sorted, pebbly sandstone rich in clay galls'. More rarely it displays mud-balls, armoured mud-balls, scattered millet-seed grains, whole sets of millet-seed grains, rolled ventifacts and a gritty nature. Invariably there is a marked erosion surface, sometimes an erosion channel, at its base. Rare sole structures, casts of footprints (*Rhynchosauroides* and *Chirotherium*) and worm burrows can be found on its undersurface. The fluviatile environment in which this facies formed involves 'the migration of megaripples in channel and point bars within the channels of rivers of low sinuosity under conditions of waning flood and net accretion'. The pebbly sandstones of the 'Delamere Member' are in places separated by, or elsewhere pass laterally into: 'Red to pink, medium- and fine-grained, well sorted, often argillaceous and micaceous, mostly trough, sometimes lenticular, planar, cross-bedded sandstone. Rare pebbles or granules are present: deformation of cross-strata and clay galls are seen'. Strata of this facies, which is more common towards the top of a rhythm, are inferred to have been the 'deposits of streams of rather higher sinuosity and lower velocity' than the streams that deposited the pebbly sandstone.

Some horizons in the 'Delamere Member' resemble a facies that is more characteristic of the Tarporley Siltstones, but which here appear to be the lateral equivalent of pebbly sandstone/'marl' in the 'Delamere Member'. The material consists of 'interbedded thin beds of pink, secondarily white, siltstone and dark red, secondarily green, micaceous shale in which the sandstone and siltstones repeatedly inject the mudstones over a thickness of 2 to 3 m'. The usual position of this lithofacies in the upper part of a sandstone-based cycle suggest that the beds are 'deposits of a river topstratum wherein the sealed but porous sand beds were in lateral hydraulic contact with the pulsating flood currents of the main river channel. Rising pore water-pressure, it is believed, led to repeated injection of sand through mudstone'.

A large part of the 'Frodsham Member' consists of sandstones similar in some ways to those that form the major constituent of the Upper Mottled Sandstone, 'but the cross beds include very large sets of lenticular planar, wedge and rarer trough types. Associated features include convex-upwards foresets, intraset cross-bedding, millet-seed lenses on foresets, a notable lack of mica and a well or very well sorted grain fabric'. The depositional environment that can be inferred from these special characteristics is an aeolian one, and the cross beds suggest that dunes may have been of dome shaped and possibly transverse types (Thompson, 1969).

The so-called 'marl bands' of the 'Keuper Sandstone Formation' are usually 'dark red micaceous shales which may show siltstone or sandstone interbeds. Mud-cracks are common'. Such strata, which in many places separate the major pebbly units of the 'Delamere Member', are interpreted as 'topstratum swale-fill and river plain deposits of streams of varying sinuosity: low sinuosity at the base of the formation: higher sinuosity towards the top. Often they are torn up by the succeeding flood-waters and end up as clay galls or mud-balls in cross-bedded sandstone'.

## PALYNOLOGY

Red-brown mudstones from the Lower Mottled Sandstone at 51.21 and 58.52 m in the Dunham-on-the-Hill Borehole proved barren of palynomorphs. At Alderley Edge, in the Stockport (98) district, a miospore assemblage dominated by *Alisporites grauvogeli* and *Protodiploxypinus spp.* but including *Densoisporites nejburgii* and regarded as early Triassic (late Scythian) in age (Warrington, 1970b) was recovered from the Alderley Conglomerate Member (Thompson, 1970b) which is there considered the basal unit of the Helsby Sandstone.

## LOWER MOTTLED SANDSTONE

The Lower Mottled Sandstone is believed to underlie some of the western marginal parts of the district, west and south of the River Dee. It is possible that in the Lache and Blacon areas, west of Chester, its upper limit lies farther west than shown on the map: boreholes in this area record 'red rock' or 'red sandstone' and give few additional details, but the occurrence of pebbles in the Blacon area is reported and these strata may fall within the Pebble Beds. Another outcrop of Lower Mottled Sandstone is inferred to occupy a fairly narrow, north-to-south belt of country immediately west of the Pebble Beds ridge of Little Barrow and Dunham-on-the-Hill. Hereabouts some 116m of sandstone underlying pebbly sandstones were pierced in the Geological Survey's Dunham-on-the-Hill Borehole. A third outcrop of Lower Mottled Sandstone almost certainly underlies ground to the east and north of the inlier of Upper Coal Measures at Edgerley and Milton Green.

Although virtually unexposed within the Chester district, the Lower Mottled Sandstone is exposed in good sections not far beyond the western margin of the district, within the adjacent Flint (108) district, and the formation is here about 305 m thick. It consists of fine- to medium-grained, soft, reddish brown sandstone with a few thin partings of 'marl' (Wedd and others, 1924, pp.138–141). Buff mottling and greenish grey layers occur, but are perhaps less common

than in the Upper Mottled Sandstone. Two of those sections were used as type sections by Hull (1869, pp.37–40). The first, in the lower part of the Lower Mottled Sandstone, is in Brad Brook [325 605], Higher Kinnerton; the other, showing the top part of the formation overlain by the Pebble Beds, is at Burton Point [303 736]. Within the Chester district, the Geological Survey's borehole at Dunham-on-the-Hill proved the upper part to be composed of mainly soft, reddish brown, medium-grained, cross-bedded sandstone. 'Millet-seed' grains are common in the more mica-free sandstones. There are thin beds and partings of chocolate silty mudstone in the upper 27 m, and the highest 11 m or so are generally finer-grained with beds of harder sandstone.

### Details

The western outcrop of the Lower Mottled Sandstone is nowhere known with certainty to be exposed. In the northernmost of the two eastern outcrops there are several roadside exposures [473 744] of red, brown and yellowish sandstone in Hapsford village, about 90–140 m north of the A.5117 road. Strahan's statement that these sandstones 'are seen to pass under the Pebble Beds' (1882, p.2) probably means no more than that they dip gently eastwards towards the outcrop of that formation.

A railway cutting [457 630] at Waverton is just within the southernmost of the two eastern outcrops of the Lower Mottled Sandstone. Here Strahan (1882, p.2) recorded soft red and white sand faulted against Pebble Beds. Farther south in the same outcrop he noted that soft red sand forms the surface near Handley Church [466 579].

The Dunham-on-the-Hill Borehole [468 715] cored 116 m of non-pebbly sandstones beneath Pebble Beds. Such a thickness suggests that these sandstones may belong to the upper part of the Lower Mottled Sandstone, and they are provisionally classified as such. Further details are given in Appendix 1, pp. 99–100.

### PEBBLE BEDS

Pebble Beds underlie extensive areas between the western margin of the district and the villages of Dunham-on-the-Hill, Tarvin, Huxley and Tattenhall. Except for one or two small sandstone exposures believed to belong to the Lower Mottled Sandstone, all the solid sections in the above area are in the Pebble Beds. The formation crops out along two eastwardly dipping north-to-south belts apparently separated by a major line of fracture that throws down west. The outcrop of the Pebble Beds along the western belt is very broad in the northwest and around Chester, but tapers southwards to a width of about 2 km at Churton. The eastern outcrop is very broad between Cotton Edmunds and Duddon, but is less than 2 km wide northwards to Dunham-on-the-Hill and southwards to Chowley.

The Pebble Beds consist of medium- to coarse-grained, reddish brown, cross-bedded sandstone, much of it incorporating characteristic rounded pebbles. The sandstone varies from incoherent to well cemented, and the latter is in places compact and durable enough to have been used as a building stone. There are also beds of sandstone with few or no pebbles, thin bands of biscuit-coloured argillaceous sandstone, and thin layers of chocolate mudstone in a few places.

**Plate 1**   Conglomerate in Pebble Beds, Dunham-on-the-Hill

'Millet-seed' sand grains are common in the coarser; more mica-free beds of sandstone. The pebbles are mostly smooth and well rounded and rarely exceed 15 cm in length, the majority being 5 to 8 cm across. They are mostly brown, reddish or grey quartzites, though some are vein-quartzes, and a few are sandstones, cherts and igneous and metamorphic rocks. In places the pebbly layers include clay-galls, while slump structures and wedge bedding are visible in some exposures. The pebbles occur either scattered thoughout the rock or locally in sufficient abundance to form lenses of shingle. Low but well-defined escarpments at Barrow Hill and Dunham-on-the-Hill are formed by about 50 m of exceptionally pebbly sandstones, the outcrop of which is shown on the map as conglomeratic sandstone (Plate 1). This local unit is belived to lie near the base of the Pebble Beds (Figure 5). Well cemented sandstones with scattered pebbles were formerly much quarried for building stone in Chester as well as in the villages of Eccleston, Christleton, Waverton, Saighton, Handley, Tattenhall and Tarvin.

The base of the Pebble Beds is nowhere exposed, but was well defined in the Dunham-on-the-Hill Borehole. That it is probably fairly well defined elsewhere is suggested by the section at Burton Point (Sheet 108) described by Lamplugh and Simmons (*in* Wedd and others, 1924, p.139). Its upper limit is, however, probably ill defined as in the Clotton Borehole and in the adjacent Wirral area to the north-west (Wedd and others, 1923, p.73). The thickness of the formation may well approach 305 m for it exceeds this figure in nearby parts of south Lancashire. Minimum thicknesses of 151m at Thornton-le-Moors and of 171 m at Mouldsworth have been proved in boreholes.

## Details

### The western outcrop

North of Chester there are many small exposures of the Pebble Beds between Backford and the Zoological Gardens, on the slopes of the late-glacial 'Deva Spillway', a broad channel now occupied by the Shropshire Union Canal. The best section is in a quarry [405 710] in Collinge Wood, some 500 m east of the A41 road bridge (Backford Bridge) over the canal, where more than 8 m of fairly hard, cross-bedded, red-brown sandstone with scattered pebbles dip gently eastwards. All other exposures are of a metre or so of sandstone usually with a few scattered pebbles. In places, however, as in the brook south of the Dale, just east of where it flows under the railway [400 697], and in an old quarry [400 710] immediately north of Backford Bridge the sandstones include lenses of shingle.

Nearer to Chester the best section is in Bache Brook, which occupies a deep ravine [401 680 to 403 581] south of the grounds of the Deva Hospital. The section begins at the mouth of the ravine south-west of Bache Hall and ends where the brook flows under the A5116. The strata dip due east at 5–10° and the following sequence through some 53 m of pebbly sandstones is exposed:

|  | Thickness m |
|---|---|
| Sandstone, soft, red, cross-bedded, pebbles common, the lower part shows contemporaneous disturbance | 4.6 |
| Sandstone, soft, red, cross-bedded, scattered pebbles | 18.3 |
| Sandstone, soft, red, cross-bedded, pebbles very scattered | 24.4 |
| Sandstone, soft, red, cross-bedded, with numerous pebbles | 1.5 |
| Sandstone, soft, red, cross-bedded, scattered pebbles | 4.6 |

East of the A5116 the valley is much shallower and is somewhat obscured by made ground.

Immediately north of Chester there is a fine artificial section along the main railway line to North Wales [401 666 to 406 671] in rock cuttings that lead into two twin tunnels. The western pair of tunnels are about 200 m long and pass beneath the Upper Northgate Street traffic interchange; the eastern pair are about 100 m long and pass beneath the precinct of Northgate Stadium (the former site of Northgate Station), alongside the Shropshire Union Canal [401 666 to 406 667] where it deeply undercuts the north wall of the city. These sections, in canal and railway, expose the same parts of the Pebble Beds sequence to a thickness of about 67 to 76 m.

The lowest beds in the railway cutting, west of the 200 m tunnels, are cross-bedded sandstones with pebbles up to a maximum of 7.5 cm long, their longer axes lying parallel to the apparent dip of the succeeding sandstones. Higher beds are also visible in a wide rock-cutting between the Upper Northgate Street traffic interchange and Victoria Road, which shows thickly cross-bedded red sandstone with creamy bands, rare pebbles and thin flaggy sandstone or sandy mudstone partings, with a true dip of about 9° to the east. Next in upward sequence are the sandstones pierced by the 100 m tunnels. The youngest beds are exposed in the south wall of the cutting immediately north-east of these tunnels. They include 3.3 m of thinly bedded, micaceous, sandy mudstone overlain by 2.4 m of cross-bedded red sandstone. The mudstone almost certainly lies within the Pebble Beds even though the section does not continue eastwards far enough to expose any overlying pebbly sandstones.

The canal section shows about 67 m of markedly cross-bedded, red, open textured, friable sandstone with rounded and sub-rounded grains and, in the lower part, only scattered small pebbles. Beds of sandstone with quartz and quartzite pebbles lie near the top of the section, notably in the cliff [405 667] 30 to 35 m west of Phoenix Tower (King Charles' Tower) at the north-east corner of the city wall. There are several well defined bedding planes which show the easterly dip to vary from 10° to 15°.

Within the walled city, red sandstone with scattered pebbles forms the foundation of buildings in Lower Bridge Street; it also underlies notable antiquities in various basements, such as the column-bases of the legionary Principia in Northgate Row, and the hypocaust known as the 'Roman Bath' on the east side of Bridge Street. Similar sandstone can be seen at intervals between Lower Bridge Street and Grosvenor Park some 550 m to the ENE. It is possible that it was the difficulty of excavating deep basements in hard Pebble Beds lying so close to the surface in most of the walled city area that led the mediaeval builders to construct semi-basements with raised 'ground floors', from which evolved the well known 'Rows' that now line the main thoroughfares. Strahan (1882, p.3) stated that 'A great portion of the stone used in the

cathedral and the old Abbey buildings adjoining it was obtained from quarries between Northgate Street and Windmill Lane and in places within the walls on the east side of Northgate Street'. He noted further that 'At several places the dressed faces of the rock indicate that it has been quarried'. I am informed that an old pit filled with quarry spoil was found in making additions to the gaol and an excavation resembling a quarry, but below the level of the river, was met with in placing the foundations of the north end of Grosvenor Bridge. It is probable that many of the quarries in Chester are of great antiquity, as this stone was used by the Romans in portions of their buildings.

On the south side of the River Dee the banks expose intermittent sections in the Pebble Beds from just east of Grosvenor Bridge to the Queen's Park area. The lowest beds crop out east of the bridge near an old mortuary chapel [403 655] in the north-west corner of the older part of Chester Cemetery, where a metre or so of red and creamy-banded sandstone with scattered pebbles can be seen. This is followed eastwards by sandstone that shows well developed slump-bedding in a cliff [404 655] some 210 m east of Grosvenor Bridge. About 137 m farther east pebbly sandstone is exposed [405 655] in the banks 90 m NW of the Rectory School and, still farther east, conspicuous wedge-bedding is a feature of higher river bank exposures [406 656] some 100 m west of Dee Bridge. East of Dee Bridge there are intermittent river bank exposures in the Queen's Park area of even higher sandstones, dipping gently eastwards as far as about 220 m east of the Suspension Bridge.

A short distance south of the River Dee cross-bedded sandstone is visible [404 652] along Overleigh Road and, a little farther south, it was seen in shallow excavations [405 651] in the vicinity of the schools.

Between Chester and Eccleston, in the west bank of the River Dee at Heronbridge [411 642], some 6 m of sandstone, carrying pebbles 5 to 7 cm long are exposed. At Eccleston, immediately west of the village, there is a good section near the bridge carrying the 'Chester Approach' to Eaton Hall over Rake Lane [408 625]. A traverse from a few metres west of the bridge eastwards towards the village shows, firstly, somewhat flaggy sandstone with fairly coarse subrounded grains, open texture, and many quartz and quartzite pebbles. Then, between 55 and 70 m east of the bridge, three parallel fractures, trending a little east of north and throwing down east, mark the position of a fault belt of unknown magnitude. East of the fractures, cross-bedded, cream open-textured sandstone with subrounded to subangular grains and only rare, small quartzite pebbles can be seen in the road side. Some 6 m of similar sandstone, lying nearly horizontally, are exposed in a nearby quarry [410 624] near Eccleston Paddocks. Nearer to the river, the steep slopes that flank the Dee alluvium from the village to Eccleston Ferry and Eccleston Lodge also offer small exposures of red, rather flaggy sandstone. Pebbles were not seen, but the exposures are considered to lie within the Pebble Beds.                              JRE, DM

South of Eaton Hall and to the east of the River Dee, the Pebble Beds form a broad drift-free ridge from Aldford to Churton. A quarry [417 588], on the west side of the road from Aldford village to Townfield Lanes, exposes about 15 m of massive, cross-bedded, brick-red, micaceous, medium-grained sandstone with quartzite pebbles in the upper part. The sandstone is strongly jointed with N–S and E–W joints, and dips almost due east at 17°. In roadside exposures [420 585] on the Aldford-Churton (B5130) road, 1.5 m of rather shaly, cross-bedded, medium-grained, brick-red sandstone with a few pebbles are visible; and 1 m of similar sandstone, rather higher in the sequence, is exposed in a lane [422 586] about 180 m to the north-east. At Ford Lane [423 581] about 1 m of shaly, red, cross-bedded sandstone, overlain by sandy clay, dips to the east at 6° to 12°. South of Aldford Hall, an elongate quarry [418 580] on the east side of the B5130 exposes 2 m of red, shaly, cross-bedded, pebbly sandstone with apparent dips of 20° to the north-east. Between this quarry and Churton there are few exposures, but a metre or so of red sandstone is visible at Grange Farm [421 573], where a well was sunk to 28 m in sandstone.

There are many exposures of Pebble Beds in and around Churton village. A quarry [417 567] at Stannage Farm exposes about 18 m of massive, cross-bedded, medium-grained, red sandstone, with scattered quartzite pebbles, dipping almost due east at 20°. Some 320 m farther south, a smaller excavation [416 563] exposes up to 9 m of similar red pebbly sandstone with a few brown shaly bands. The higher beds at Churton are rather shaly, and are well exposed east of the B5130, in a lane between the inn and Church House Farm [422 565].

South of Churton, and only about 1 km beyond the southern margin of the district, lies a fine section exposing about 160 m of Pebble Beds in the east bank of the River Dee on both sides of Farndon Bridge. Details of the section are given by Poole and Whiteman (1966, p. 24).                                              EGP

In the more easterly parts of the western outcrop there are several drift-free areas of Pebble Beds from Hoole Bank in the north, through Christleton and Waverton, to Saighton in the south. Pebbly and cross-bedded sandstones of the Hoole Bank outcrop are exposed in a lane that runs to the south and south-east of the mansion [430 692]. They were also bared in shallow cuttings [433 687] for the entrance to the car park of the Royal Oak public house on the south-east side of the A56 road in Hoole village. In a cutting [434 684] of the disused railway immediately south of Hoole village, Strahan (1882, p. 2) recorded, beneath a bed of sand, soft red, white and yellow sandstones which he thought belonged to the Lower Mottled Sandstone. In view of the very close proximity of the Royal Oak exposure of Pebble Beds it is more probable that it is this formation that crops out in the railway cutting.           JRE

At Christleton, a quarry [438 655] alongside the Shropshire Union Canal, exposes loosely cemented, cross-bedded, red sandstone with subangular to subrounded grains, and some cream-coloured sandstone bands. Although no pebbles were seen in the faces of the quarry, a few occur in the locally derived building stone used in constructing the adjacent houses. Observed dips of 8° to 10° to the east are considered to be true dips. Small scattered exposures of similar sandstone occur in Christleton village.

At Waverton, the Pebble Beds are exposed dipping eastwards at 8° to 9° in the partially flooded Waverton Quarry [455 633], as well as in an adjacent railway cutting near the A41 road bridge. Pebbles are generally rare in the main body of the sandstone, but they are more common in a bed near the top of the south-east face of the quarry. It was in the railway cutting a short distance south-east of the road bridge that Strahan (1882, p. 2) saw Pebble Beds faulted against Lower Mottled Sandstone. His interpretation has been accepted in constructing the current maps.

The village of Saighton is largely built on drift-free Pebble Beds, the best exposures of which are in the road [422 618] by Saighton Grange on the south side of the village. The lowest beds are visible on the west wide of the road; they consist of loosely cemented sandstone with rounded and subrounded grains and include a few thin bands of breccia-conglomerate. These pass up into pebbly sandstones which are well exposed along the east side of the road, dipping eastwards at about 16°.                           DM

*The eastern outcrop*

At Dunham-on-the-Hill there are many roadside sections in conglomeratic sandstones believed to form the lowest strata of the Pebble Beds in this part of the district (Plate 1). The lowest exposures are in a small quarry [470 725] on the east side of the A56, 200 m SSW of the Dunham Arms Inn, where the following strata dip a little south of east at 15°: fairly hard, red-brown, cross-bedded, pebbly sandstone 4.5 m; overlain by soft, coarse, red-brown sandstone 1.5 m; then fairly hard, red-brown, cross-bedded, pebbly sandstone 3 m. Slightly higher sandstones with shingly lenses are exposed in

the lower part of the lane that ascends from near the Dunham Arms up to the middle of the village. All along the higher, middle part of the village, and especially in the lanes near Town Farm [471 724] there are exposures, amounting to more than 30 m of medium- to coarse-grained, reddish brown cross-bedded sandstone with scattered pebbles and lenses of shingly conglomerate. The dip is consistently a little south of east at 10° to 15°.

At Little Barrow there is a good section in the railway cutting, beginning at the former railway station [468 702], just west of the B5132 road bridge, and extending for 300 m east of the bridge. From the former south platform west of the bridge to the rock face immediately east of the bridge, the following ascending sequence is visible: soft, red, cross-bedded sandstone, about 1 m; overlain by massive, red-brown sandstone with scattered pebbles, 1.75m; then lenticular pebbly conglomerate and sandstone, 0.5 to 1.5 m; followed by massive, red-brown sandstone with scattered pebbles. Eastwards along the cutting these sandstones are overlain by thickly bedded, pebbly sandstones that dip eastwards at about 7° and are estimated to be about 15 m thick. The easternmost 155 m of exposed cutting show soft, cross-bedded, red sandstones with scattered pebbles, but with a single pebbly lens near the middle; all the strata dip gently eastwards. The largest pebble seen in the section was 13 cm long.

The same conglomeratic sandstones make the main ridge of Little Barrow and of Barrow Hill to the west of the B5132. There are many small exposures of them in Little Barrow village, while hard beds of conglomerate make distinct features along the western slopes of the hill 0.5 to 1 km SSW of the roadbridge. A little farther south, at Great Barrow, sandstones with only a few pebbles lie nearly horizontal, and their relationship to the conglomeratic sandstones of the higher ground west of the village is interpreted as a faulted one. The conglomeratic sandstones are believed to be low in the Pebble Beds, whereas the sandstones at Great Barrow church and Barrowmore Hospital are probably higher in the sequence, possibly not far below the Upper Mottled Sandstone. More than 6 m of nearly horizontal, red-brown, cross-bedded sandstones with softer silty layers are exposed in the road adjacent to, and in the driveways into, Barrow church and rectory [470 684]. Similar strata are also seen near the middle of the village where, again, they are nearly horizontal. At, and immediately west of the western entrance to the grounds of Barrowmore Hospital, a road cutting [474 691] exposes 7.5 m of massive, and cross-bedded, soft, red sandstone, overlain by 1.5 m of soft, red, laminated sandstone with thin grey layers, followed by 4.5 m of massive, cross-bedded, soft red sandstone containing rare isolated pebbles. The dip is very gently eastwards and the highest beds in this section are exposed intermittently for some 600 m along a lane which follows the strike NNE from the hospital entrance to Broomhill.              JRE

Around Tarvin and Hockenhull outcrops of pink-red, cross-bedded, pebbly sandstone are marked by gently rounded features. In Tarvin itself the sandstone can be seen at several places adjacent to the A51 road through the village and, west of the village, red sandstone with pebbles is visible [487 670] alongside the A54 bypass, east of Grove House. Similar sandstone is exposed in a small excavation [491 669] near Ducker's Well, about 80 m south-west of the church. South-east of Hockenhull Hall, pebbly sandstone, apparently horizontal, crops out [486 660] in the side of the lane; almost horizontal flaggy sandstone is exposed [479 664] in a small stream SSW of Holme Street Hall. South-west of Hockenhull Hall an exposure [489 657] in Platts Lane, the old mediaeval road to Chester across the River Gowy, shows sandstone with a few quartzite pebbles, apparently dipping at 10° in a direction slightly south of west. Further exposures occur near Round House [495 649] Stapleford, and along Waterless Brook [592 638] 600 to 700 m west of Burton Hall.                    DM, JRE

South of Huxley, in disused quarries [509 604] by the Shropshire Union Canal aqueduct over the River Gowy, 5 m of brick-red, coarse, markedly cross-bedded conglomerate and pebbly sandstone overlie 4 m of red, fine-grained, non-pebbly, rather argillaceous sandstone with coarsely micaceous shaly bands. About 90 m west of the main quarry 4.5 m of the lower sandstone are exposed in old sandpits. Some 1 m is also visible to the south, in the bank of the River Gowy at Brook Hole [509 603]. In the quarries, the base of the conglomerate is sharp and slightly erosive; its lower part is well cemented but becomes more shaly upwards and passes into massive sandstone with scattered large, rounded quartzite pebbles. The dip is to the east at 10°.

Near Tattenhall Smithy and mill pond [487 583] about 4.5 m of soft, red, cross-bedded sandstone with scattered pebbles have an apparent dip of 10° to the SSE. Farther south-west, in Rocky Lane, the cutting at The Righi [481 577] exposes up to 2.5 m of rather shaly, pebbly sandstone, overlain by thin boulder clay, and dipping due south at 7°. At Handley, in old quarries [471 575] alongside the A41 near the smithy, some 21 m of rather shaly, fine- to medium-grained, cross-bedded, pebbly sandstone dip south-eastwards at 17°. These strata, dipping at a relatively high angle away from the nearby inlier of Coal Measures, probably lie low in the Pebble Beds, and soft red sand seen by Strahan (1882, p. 2) near Handley Church may be the Lower Mottled Sandstone.

At Chowley, on the southernmost part of the eastern outcrop, some 14 m of massive, cross-bedded, pebbly sandstone are visible in Dog Lane [477 563] on the east side of the former railway bridge. The dip is to the east at about 10°; just east of Chowley Oak Lane the uppermost beds are shaly, with small pebbles of red mudstone and bands of coarse, red 'millet-seed' sandstone.          EGP.

BOREHOLE INFORMATION  Pebble Beds have been pierced in many boreholes, most of them for water supply. A notable series was drilled into the northern part of the western outcrop, between Little Stanney and Thornton-le-Moors, for the Shell Refining and Marketing Company. One of these, 'Shell No.5' [441 748], situated immediately north-west of Thornton village, started on an almost drift-free outcrop of Pebble Beds and cored Pebble Beds to a depth of 152 m. Sections of No.2 and No.19 boreholes, west of Thornton-le-Moors, are given in Appendix 1, p. 107.

Geological Survey boreholes at Dunham-on-the-Hill [468 715], Cotton Hall [465 662] and Clotton [532 637] were drilled during the primary six-inch survey to define the eastern outcrop of the Pebble Beds. The Dunham-on-the-Hill Borehole (Appendix 1, p. 99) cored what is believed to be the basal 9 m of the Pebble Beds, and the Cotton Hall Borehole (Appendix 1, p. 98) cored 8 m of the lower part of the formation. The Clotton Borehole (Appendix 1, p. 97) cored the uppermost 63 m of the Pebble Beds and the cores revealed a very gradual passage from Pebble Beds into Upper Mottled Sandstone.

Two boreholes at Mouldsworth Pumping Station [503 704] reach a depth of 307 m. The driller's log of No.1 Borehole records about 76 m of drift on sandstones, which are pebbly from 186 m to 245 m and from 272 m to 307 m; thus at least 121 m of Pebble Beds were cored. T.A. Jones (1935, p. 276) classed the whole 231 m of sandstones pierced by the boreholes as Pebble Beds, though he stated that pebbles were first noticed at a depth of 136 m: on this basis the minimum thickness of the Pebble Beds is 171 m. Jones did not explain why he classified the uppermost 60 m of sandstones, which presumably lack pebbles, as Pebble Beds, and because there are nearby outcrops of sandstones known to be high in the Upper Mottled Sandstone, the higher sandstones at Mouldsworth Pumping Station are here provisionally referred to the Upper Mottled Sandstone. It is not possible, however, to be sure exactly where the Pebble Beds/Upper Mottled Sandstone boundary should be drawn.

A similar problem, here caused by the sporadic occurrence of pebbles much higher in the sequence than the highest acceptable formational boundary, is illustrated by some nearby boreholes at Ashton. Two of these [504 688 and 507 689] cored, below about 45 m of drift, some 50 m of sandstone with a very few scattered

quartz pebbles underlain by at least 151 m of non-pebbly sandstones that are almost certainly referable to the Upper Mottled Sandstone.

An even more interesting borehole, Willington No. 1, sited [531 653] about 700 m south of Willington Hall (Appendix 1, 108–109) on what was believed to be basal Helsby Sandstone close to the north-east (downthrow) side of a fault-belt, actually started well down in the Upper Mottled Sandstone, and presumably on the upthrow side of the fault-belt. Beneath about 24 m of drift it cored soft red micaceous sandstone with siltstone bands and some mudstone galls, passing down into soft red sandstone with siliceous pebbles. The first pebble was at a depth of 198 m, below which they became gradually larger and more abundant to the bottom of the borehole at 305.5 m. In all about 107 m of the upper part of the Pebble Beds were cored.          JRE

## UPPER MOTTLED SANDSTONE

The main outcrop of the Upper Mottled Sandstone forms a relatively narrow belt separating the eastern outcrop of the Pebble Beds from the outcrops of the Helsby Sandstone and Tarporley Siltstones. It extends from Helsby southwards beneath Ashton, Clotton and Brassey Green to Burwardsley and Bolesworth in the extreme south. There is also a narrow outcrop east of the Helsby-Manley outlier, and another east of the Peckforton Hills at Peckforton. The small outcrop depicted west of Thornton-le-Moors is of doubtful validity.

Unlike the Lower Mottled Sandstone, the Upper Mottled Sandstone of this district, particularly the upper part, is visible in many natural exposures, and has been examined in many borehole cores. It consists of soft or incoherent, fine- to medium-grained, cross-bedded sandstone, red and reddish brown in colour, but extensively mottled with buff and grey patches. This character, from which the name 'Mottled Sandstone' was originally derived, is probably of secondary origin, being attributable to post-depositional processes possibly associated with mineralisation. There are also sporadic thin beds of grey sandstone, and partings of red or grey silty shale in which the grey colour may or may not be original. The coarser, more mica-free beds of sandstone usually contain 'millet-seed' sand grains, and there are bands of laminated, almost shaly sandstone at some levels: the latter tend to be recorded as 'marl' in drillers' logs. A summary of the sedimentological aspects of the cross-bedding and related features as analysed by Thompson (1970a) has already been given (p. 15).

The base of the Upper Mottled Sandstone is everywhere transitional, and this fact, together with the poor exposure of the transitional strata, means that there is no section within the district where the base can be accurately drawn at outcrop. Those boreholes which start in Upper Mottled Sandstone and pass into Pebble Beds show a passage from one formation to the other, and the scattered occurrence of pebbles in the highest part of the Pebble Beds means that, even in cores, a precise boundary between the two formations is hard to define. The top of the Upper Mottled Sandstone is, in some places, sharply defined by the slightly erosive base of the lowest clay-gall conglomerate or pebbly sandstone of the Helsby Sandstone where this horizon locally coincides with the major regional change in the lithofacies of the sandstone. Elsewhere beds of typical Upper Mottled Sandstone facies interdigitate with harder sandstones containing clay-gall

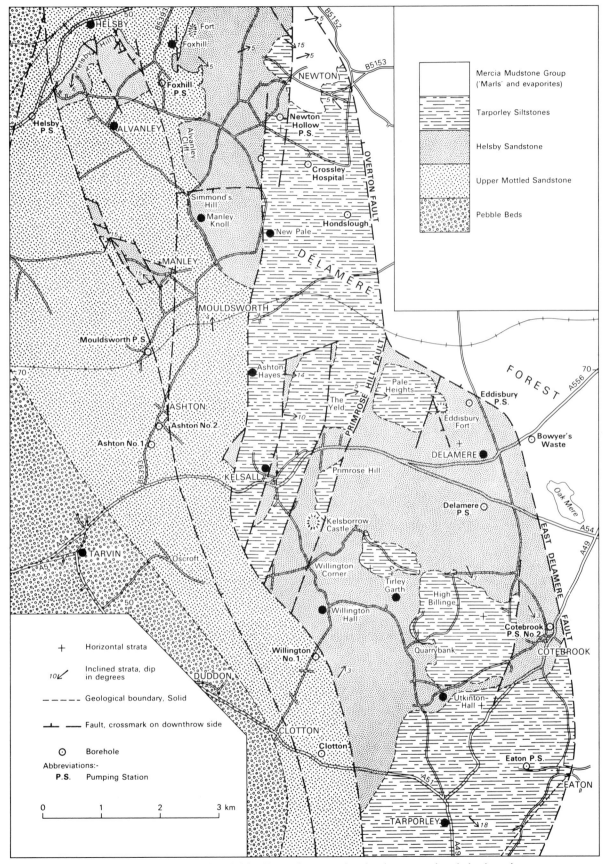

**Figure 6** Map of the solid geology of the Helsby-Tarporley area showing important boreholes into the Helsby Sandstone and associated strata

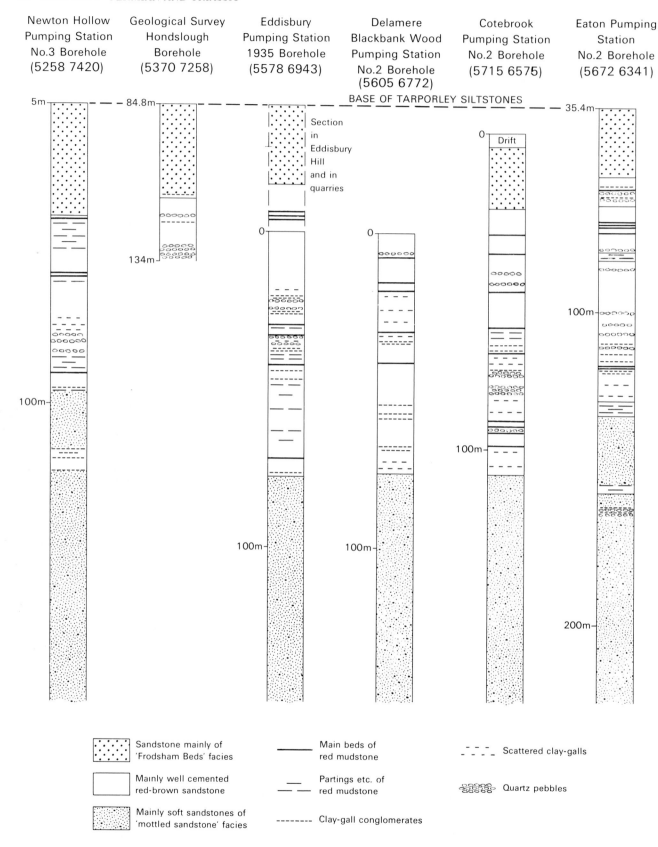

**Figure 7** Comparative generalised sections of cored boreholes in the Helsby Sandstone and associated strata between Newton and Eaton

conglomerates, before the onset of the major pebbly sand-stone facies of the Helsby Sandstone. In such places the top of the formation is not as easily defined. Definition is made no easier by recognising the zone of interdigitation as a separable unit (e.g. Thurstaston Member of Thompson, 1970b) and, consequently, this has not been attempted. In most places the top of the Upper Mottled Sandstone has been taken at the incoming of clay-gall conglomerates and pebbly sandstones that are more than thin, isolated bands. Defined thus the boundary is somewhat subjective and may be slight-ly transgressive within the district. Even so, it is as reliable for mapping purposes as most other stratigraphical boun-daries within the Trias. It is discussed further in connection with the thickness variations of the Helsby Sandstone (see p. 26 and Figures 6 and 7).

The thickness of the Upper Mottled Sandstone probably exceeds 305 m throughout the district. About 274 m were proved in a borehole which starts at least 46 m below the top of the formation at Foxhill Pumping Station [506 748] on the northern margin of the district. Nearly twenty kilometres to the south, just beyond the southern margin of the district, about 305 m were proved in a borehole [528 550] which starts some 76 m below the top of the formation at Peckfor-ton Gap Pumping Station.

## Details

### The north

The lowest part of the Upper Mottled Sandstone is unexposed in the north, and only part of the middle of the formation is exposed in the railway cutting and in quarries near the former railway station at Mouldsworth. Some of the higher sandstones may be seen in lanes near Manor Farm, Manley. Fine exposures of the upper part of the Upper Mottled Sandstone around the lower slopes of Helsby Hill lie immediately north of the northern margin of the district in the adjacent Runcorn (97) district. The best section is in the rock-cut lane (Rake Lane) that ascends south-eastwards from the A56 at the Railway Inn [490 757], then swings southwards as Old Chester Road and Alvanley Road climb for 600 m or so to a sharp elbow-bend [490 750] above the former Helsby Quarry (see Figure 8).

The railway cutting [513 708] about 90 m east of the former Mouldsworth railway station (east side of the B5393 road bridge) shows 6 m of soft, red and yellow mottled, cross-bedded sandstone. More than 21 m of similar strata have been dug for sand in the adja-cent quarry of the Ship Canal Sand Company [514 709]. A N-S fault with an eastward downthrow crosses this quarry, and there are also several smaller north-to-south fractures and shear-planes. The dip is due north at 4°. About 7.5 m of somewhat harder and mainly grey, cross-bedded sandstones are seen in the small quarry just east of the B5393 about 90 m north of the railway. Faulting makes it dif-ficult to be sure where these exposures lie within the Upper Mottled Sandstone.

Sandstones high in the Upper Mottled Sandstone are exposed at Manley (see Figure 8) in a lane from Manor Farm to Lower Hall Farm [497 721], and in the lane and cutting for the old railway siding on the west side of Manley Quarry [498 719]. All these ex-posures are of soft, cross-bedded, red and yellow mottled sandstone.

The northernmost part of the main outcrop of the Upper Mottled Sandstone was pierced by boreholes at the British Insulated Callender Cables Ltd [482 746] and the former Helsby Waterworks [484 743]. In the Cable Works borehole, 106 m of Upper Mottled Sandstone were pierced beneath 46 m of drift. The outcrop east of the Helsby-Manley outlier (see Figure 8) was penetrated by three

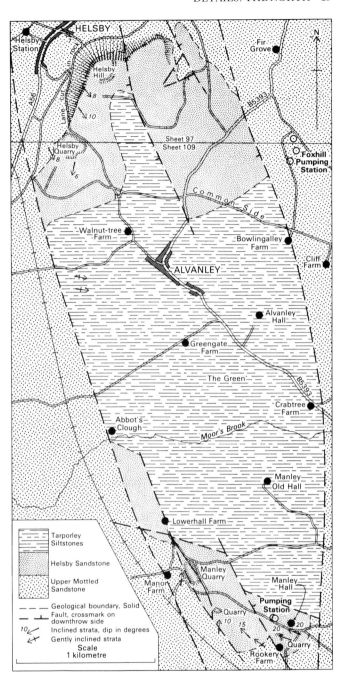

**Figure 8** Map of the Helsby-Manley outlier showing the Helsby Sandstone and associated strata

boreholes at Foxhill Pumping Station [506 748], all of which were drilled entirely in Upper Mottled Sandstone (Appendix 1, p.101). No. 2 Borehole is the deepest and proved the formation to a depth of 284 m beneath 9.5 m of drift. East of this outcrop, at Newton Hollow Pumping Station [526 743], three boreholes all passed into Upper Mottled Sandstone beneath a thick cover of Helsby Sand-stone. The deepest, No. 3 Borehole, reached the base of the Helsby Sandstone at a depth of about 96 m and proved Upper Mottled Sandstone to 305 m. Although fully cored, much of the sandstone was so soft that it disintegrated on extraction. South of Newton Hollow, on Birch Hill, a borehole [522 734] at the hospital (former-ly Liverpool Sanatorium) reached the base of the Helsby Sandstone at a depth of about 88 m and pierced 36 m of Upper Mottled Sand-stone.

South of the Simmond's Hill escarpment several boreholes at Manley Common Pumping Station [515 717] pierced the Upper Mottled Sandstone beneath about 11 m of drift, and continued in the formation, some of which was very soft, to a depth of 183 m. Inadequate records of the boreholes to 183 m at Manley Quarry Pumping Station [490 719], 1 km west of the quarry, make any interpretation of them speculative.

The interpretation of the boreholes at Mouldsworth Pumping Station [503 704] north of Ashton has already been discussed (p. 20). More-recent boreholes just south of Ashton [504 688 and 507 609] cored sandstones mainly of typical 'mottled sandstone' facies to depths of 238 m and 152 m beneath some 44–46 m of drift. A very few scattered quartz pebbles occurred in the uppermost 50 m or so of strata cored, but otherwise the sandstones appear to be of typical Upper Mottled Sandstone.

On the eastern slopes of the Delamere Hills, a borehole (Appendix 1, p. 101) at Eddisbury Pumping Station [558 695] reached the base of the Helsby Sandstone at a depth of about 78 m, cored Upper Mottled Sandstone to a depth of 153 m, and was later drilled deeper in the same formation to 229 m. South of Delamere, at the Delamere (Blackbank Wood) Pumping Station [560 677] a borehole (Appendix 1, p. 00) reached the base of the Helsby Sandstone at about 76 m and cored Upper Mottled Sandstone to 244 m. In these boreholes the lithology of the formation is typical, and much of the sandstone core was so soft that it disintegrated on extraction.　JRE

*The central area*

There are no exposures of undoubted Upper Mottled Sandstone in the tract between the A54 and the River Gowy, and the Geological Survey's Clotton Borehole [532 637] was drilled to elucidate some of the problems of a critical part of this area. The borehole (Appendix 1, p. 97) cored the lowest 43 m of the Upper Mottled Sandstone beneath 8 m of drift. It illustrated how the pebble content of the highest Pebble Beds diminishes upwards so gradually that even the presence of 2 or 3 pebbles in some 10 m of sandstone tends to be used as the main criterion by which such sandstones are excluded from the Upper Mottled Sandstone. The more recent Willington Lane Borehole [531 653] in the problematical Clotton-Willington ground pierced, beneath 24 m of drift, about 174 m of soft red micaceous sandstone with siltstone bands and mudstone galls, underlain by sandstones with rare quartz pebbles (Appendix 1, p.108).

East of the outcrop, a borehole (Appendix 1, p. 98) at Cotebrook Pumping Station [572 657] entered Upper Mottled Sandstone at a depth of 108 m and cored the highest 136 m of the formation. Similarly a borehole (Appendix 1, p.100) at Eaton Pumping Station [567 634] entered Upper Mottled Sandstone (the Thurstaston Member of Thompson) at a depth of about 133 m and cored the uppermost 111 m. The Eaton Borehole is noteworthy in that there is a thin development of clay-gall conglomerate with quartz pebbles at about 29–32 m below the top of the Upper Mottled Sandstone.　JRE

*The south*

Only the upper part of the Upper Mottled Sandstone is exposed on the flanks of the hills around Beeston. Typical lithology is seen in the Grotto [539 590] 350 m south-east of the keep of Beeston Castle (see Figure 9), where about 7.5 m of friable, yellow-leached, fine- to medium-grained, markedly cross-bedded sandstone dip E11°S at 11° (Plate 2). Bands of very coarse 'millet-seed' sand occur, and harder, barite-rich veins and patches are prominent on weathered faces. West of the Grotto, about 2.5 m of soft red-brown shaly sandstone with fawn-leached bands, lying somewhat lower in the sequence, dip E11°S at 7° in an old adit mouth [538 590], now partially blocked. A further exposure [539 588] of 2.5 m of rather shaly red-brown sandstone is seen dipping E27°S at a locally high dip,

this being due to the proximity of the north-trending Peckforton Fault.

In a gully immediately north-west of Beeston Castle Lodge [539 591], about 30 cm of hard Helsby Sandstone with angular grains rests on 2.5 to 3 m of soft, fine-grained, cross-bedded red sandstone with both subangular and 'millet-seed' grains, and with harder nodules and patches, probably rich in calcite or barite. This sandstone rests in turn on 2.5 m of harder coarse 'millet-seed' grained red sandstone. The easternmost exposures show undulations of dip, with strong jointing and slickensiding associated with the Peckforton Fault.

Some 1.22 m of shaly, fine- to medium-grained, pink, cross-bedded sandstone with coarse 'millet-seed' grains are exposed [532 581] in the side of a lane north-west of Peckforton Castle (see Figure 9). In an adjacent disused quarry east of the lane, about 4.5 m of similar soft pink and fawn mottled sandstone dip due east at 10°. To the south, about 1.22 m of soft fine-grained mottled sandstone are visible north-west of Stanner Nab [530 575] and 1.5 m of similar bright red and fawn mottled sandstone dip in a southerly direction in an exposure [529 571] SSW of Stanner Nab. In a disused quarry [519 566] south of Burwardsley Hall Farm (see Figure 9) about 12 m of typical Upper Mottled Sandstone are exposed and, at the south end of Willow Hill, 3 m of shaly red mottled sandstone and sandy shale rest in a roadside exposure [518 562] on 3 m of more massive red-brown and fawn mottled sandstone. These exposures

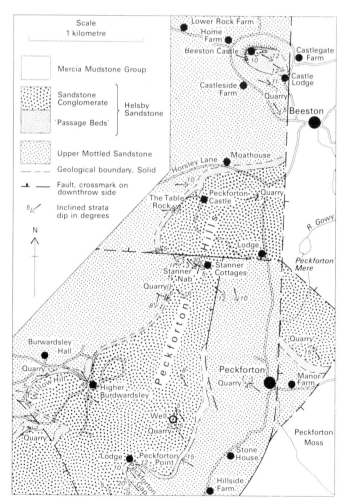

**Figure 9** Map of the Upper Mottled Sandstone and Helsby Sandstone outcrops of the Peckforton Hills and Beeston Castle

**Plate 2**   Upper Mottled Sandstone showing cross-bedding, Beeston Castle Grotto

immediately underlie the 'Passage Beds' at the base of Helsby Sandstone.

The westernmost exposures of the upper part of the Upper Mottled Sandstone lie along the foot of the scarp behind Bolesworth Castle [495 560]. Here 2.5 m of soft red and fawn mottled cross-bedded sandstone dip eastwards at 5° beneath 9 m of pebbly Helsby Sandstone.

There are good sections of the sandstones immediately underlying the Helsby Sandstone on the south-west side of the deep hollow known as Peckforton Gap (see Figure 9). Exposures [527 557] close to the southern boundary of the district show up to 12 m of Upper Mottled Sandstone, markedly cross-bedded and in places including slump structures. In the floor of Peckforton Gap, just south of the district, boreholes sunk for the Staffordshire Potteries Water Board start more than 76 m below the top of the Upper Mottled Sandstone and penetrate up to a further 305 m of the formation without reaching Pebble Beds (Poole and Whiteman, 1966, pp.29 and 111).

EGP

## HELSBY SANDSTONE

The Helsby Sandstone makes much of the high ground of the mid-Cheshire hills, but the outcrop of the formation is broken by faults and is discontinuous. There are three main areas of outcrop. One includes the northern and southern flanks of the Helsby-Manley outlier (see Figure 8), west-facing escarpments and hilly ground between Rileybank and Rangeway Bank Farm, and narrow faulted outcrops near Kelsall. The largest outcrop lies around Delamere, Willington, Quarrybank and Cotebrook, where the beds are near-horizontal or very gently inclined. Finally, in the south, Beeston Castle surmounts a small outlier of Helsby Sandstone, detached from another major outcrop that forms Peckforton Hills, and has a faulted appendage between Burwardsley and Bolesworth Castle.

At outcrop in the north, the lower part of the Helsby Sandstone consists of fairly well cemented, reddish brown, but locally grey or buff, mainly coarse-grained, cross-bedded sandstones with much clay-gall conglomerate, especially in the lower parts of each major unit. Some interstratified, dark red, micaceous, shaly mudstones or siltstones occur; some of these form well defined beds in which mud-cracks are common: some are very variable bands, partings or wisps, that divide the major beds of sandstone. The thicker units of sandstone constitute the massive building stones visible in quarry faces, and generally contain small quartz pebbles which are mostly conspicuous, though fairly scattered; some of the lower sandstones are locally non-pebbly. The upper 20 to 30 m of the Helsby Sandstone are here termed the 'Frodsham Beds' facies. They are usually soft, very highly cross-bedded, red-brown, buff or grey sandstones, not appreciably pebbly but commonly containing numerous small nodal accretions of harder, barite-cemented material.

Reference to the sedimentological significance of the cross-bedding and other features of the 'Frodsham Beds' facies has been made above (pp. 15–16). In some respects this facies resembles the Upper Mottled Sandstone, but in many places the beds are clearly more coherent and resistant to erosion than the latter. Where free from drift the more resistant parts give rise locally to bold features or elsewhere to a distinctive topography.

At Helsby Crags, immediately north of the district, the base of the Helsby Sandstone is well defined by a prominent bed of clay-gall conglomerate that rests on typical Upper Mottled Sandstone and is overlain by harder pebbly sandstones. About 43 m of the Helsby Sandstone form the crags and summit of the hill, and about as much again, probably mainly of 'Frodsham Beds' facies, occupy the less exposed hill-top slopes between the trigonometrical station and a large marl-pit in the Tarporley Siltstones some 350 m to the SSE. The total thickness at this type locality is, therefore, of the order of 85 m. The full thickness of the Helsby Sandstone was pierced in a borehole at Newton Hollow Pumping Station [526 743], ESE of Helsby. The logs of this and of two adjacent boreholes recorded no sharp lithostratigraphical change at the probable position of the base of the Helsby Sandstone. Nevertheless, the available information when plotted (Figure 7) suggests that the base does not lie higher than about 85 m below the base of the Tarporley Siltstones. Below this level about 27 m of strata include a few layers of somewhat incoherent clay-gall conglomerates near the base, and the proper assignment of these strata is uncertain. The thickness of 85 m, however, compares so closely with that determinable at outcrop at Helsby that this thickness is preferred at Newton Hollow, and may be regarded as an acceptable average for the northern part of the district.

From Helsby southwards to Eaton at the southern end of the Delamere outcrop, the base of the Helsby Sandstone has generally been taken at the incoming of pebbly and clay-gall conglomeratic sandstone as a major constituent, the larger clasts of the conglomerates being usually of red mudstone or siltstone and many of the units having quartz pebbles. Boreholes at Eddisbury [557 694] and Delamere [561 677] pumping stations (Figure 7), which start between 30 and 45 m below the top of the Helsby Sandstone, prove that part of the sandstone lying below the 'Frodsham Beds' facies to be at least 76 m thick. This makes a total thickness for the Helsby Sandstone of between 105 m and 120 m. Cores of the borehole at Eddisbury were examined in 1935 by W. B. Wright who referred the sandstones down to 78 m to the 'Keuper Sandstone'. Cores of the borehole at Delamere, examined during the primary survey, confirmed Wright's findings at Eddisbury. Thus there is a thickness variation from about 85 m at Helsby to as much as 120 m around Delamere. The persistence of the 20 to 30 m of 'Frodsham Beds' facies in the upper part of the Helsby Sandstone, and the close resemblance of the exposed middle part of the formation

around Delamere to the lower part at Helsby suggest that the increase in thickness is due to the incoming of conglomeratic sandstones south-eastwards from Helsby beneath the base of the main conglomerates there exposed. The logs of the Newton Hollow boreholes are probably to be interpreted in this light. A similar interpretation may be applied to the section of the Helsby Sandstone in a borehole at Eaton Pumping Station [568 634] where a total thickness of 129 m was included within the formation by D. A. Wray. Here, conglomerate assigned by Wray to the base of the 'Keuper Sandstone' is separated from the main part of the formation by nearly 30 m of poorly cemented sandstone, most of which is of mottled sandstone facies. This again suggests that, between Helsby and Tarporley, strata classed as the highest part of the Upper Mottled Sandstone in one place and those classed as the lowest part of the Helsby Sandstone in another place may be different facies of the same 30 m of strata.

In the Beeston-Peckforton area the lower part of the Helsby Sandstone consists of 15 m or less of thickly bedded, non-pebbly, reddish brown, coarse-grained, well-cemented building stones that have been called 'Passage Beds' (Poole and Whiteman, 1966, p. 17). These are overlain by conglomeratic sandstones up to 30 m thick that are shown separately as Sandstone Conglomerate on the map. The Sandstone Conglomerate commonly contains boulders of sandstone and fragments of red and green mudstone: it is heavily mineralised near faults, with hematite and barite. It becomes less pebbly upwards and passes up into massive, non-pebbly, reddish brown building stone. South-westwards towards Bolesworth Castle, the 'Passage Beds' disappear beneath the unconformable base of the Sandstone Conglomerate which comes to rest directly on the Upper Mottled Sandstone. The precise relationship of the Sandstone Conglomerate of this area to the several pebbly and conglomeratic levels farther north is uncertain. Thus, in the Peckforton Hills the disappearance of the basal 'Passage Beds' in a south-westerly direction, by overstep of the Sandstone Conglomerate seems to be a different type of lateral change from that which occurs in the equivalent strata in a north-westerly direction from Delamere to Helsby.

The upper part of the Helsby Sandstone in the Peckforton Hills area crops out east of Bolesworth Castle where the Sandstone Conglomerate is overlain by 73 m of incompletely exposed strata. The lowest of these include hard, coarse-grained, reddish brown sandstones, and the highest 30 m or so are of 'Frodsham Beds' facies which are abruptly succeeded by the well-bedded mudstones, siltstones and sandstones of the Tarporley Siltstones. This part of the sequence is, therefore, comparable with that at Helsby despite a certain amount of variability in the intervening ground.

## Details

*The north (Helsby to Delamere)*

Although just beyond the limits of the district a summary of the important section of the Helsby Sandstone in Helsby Crags [492 754] is given here:

|  | Thickness m |
|---|---|
| Coarse, cross-bedded, reddish brown sandstone with many quartz pebbles | 8 |
| Very coarse, soft red and buff sandstone with layers of incoherent marly sandstone | 6 |
| Coarse, cross-bedded brown and buff sandstone, the upper part in places very soft | 3.5 |
| Coarse, massive and cross-bedded brown sandstone with many quartz pebbles and several layers of clay-gall conglomerate | 1.25 |
| Medium to coarse, massive and cross-bedded brown sandstone; many layers and partings of clay-gall conglomerate; a prominent bed of clay-gall conglomerate at base | 12.5 |
| Upper Mottled Sandstone | 1 |

The above section is visible in the sides of a prominent fault gully that makes a gash in the main face of Helsby Crags. The lower part is best seen just north of the gully, and the upper part in near-vertical crags on its south side. Helsby Quarry [491 749], formerly the best section in the Helsby Sandstone along the northern margin of the district, is now filled in. It exposed almost all but the top part of the sandstones which make the high crags to the north. Strahan (1882, p.7) gives the following section:

|  | Thickness m |
|---|---|
| Sandstone, good building stone | 7.63 |
| Soft white, current-bedded sand with two 8-inch bands of dark red shale | 1.52 |
| Ditto, with irregular red seams | 3.05 |
| Dark red shale with bullions | 2.13 |
| Sandstone, good building stone | 12.19 |

At the south end of the Helsby-Manley outlier (see Figure 8), the Helsby Sandstone was worked at Manley Quarry [499 720]. Here the deepest part of the quarry at the north end exposes more than 15 m of tool-trimmed massive grey sandstone beneath a thin layer of red and grey shaly mudstone. The south-western face and the higher part of the north-eastern face of the quarry are faulted. Strahan (p.9) refers to 'the pure white building stone of Manley' and states that 'The Manley stone occurs in a bed about 45 feet thick, the best quality being at the bottom. The grain is coarse and sharp and generally too loose to make a first-class building material'. He further observes that 'the stone has been used in the Grosvenor Bridge and Chester Castle, including the monoliths supporting the portico' (presumably the massive Doric columns of Harrison's neo-classical propylaea and those fronting the law courts) 'and in parts of Eaton Hall'.

Scarred ridges of Helsby Sandstone dipping north-westwards make much of the wooded ground south-east of Manley Quarry and there is a short section in the road [506 716] by Manley Hall (see Figure 8):

|  | Thickness m |
|---|---|
| Mainly massive brown sandstone with scattered quartz pebbles and clay-gall cavities | 3 |
| Massive and wedge-bedded brown sandstone with softer lenticular layers | 4 |
| Massive and cross-bedded brown sandstone, a few small pebbles; top and base defined by true bedding planes | 3.5 |
| Very coarse, fairly soft, pebbly sandstone | 2 |
| Cross-bedded, reddish brown sandstone with several softer very coarse lenticular layers, some with galls of red shaly mudstone | 8 |

There are many small exposures of Helsby Sandstone along the main escarpment between Rileybank and Rangeway Bank Farm. The best section is that in the bold crags on the western and northern slopes of Simmond's Hill [511 728]. The following section is visible:

|  | Thickness m |
|---|---|
| Massive and cross-bedded, brown sandstone; the lower 1.5 m or so very coarse with quartz pebbles | 3 |
| Soft, coarse, reddish brown sandstone with abundant galls of red mudstone; lenticular | 1 |
| Massive and cross-bedded brown sandstone with scattered red mudstone galls or cavities | 3 |
| Soft, coarse, buff and red sandstone with abundant galls of red mudstone; lenticular | 0.5 |
| Coarse, cross-bedded, reddish brown sandstone soft sandy parting with red mudstone galls | 3 |
| Cross-bedded buff sandstone soft red mudstone parting, impersistent | 1.5 |
| Massive and cross-bedded buff sandstone soft red mudstone parting | 3 |
| Massive brown sandstone, scattered clay-gall cavities | 2 |

The highest sandstones of the above section were worked in quarries [512 727] on the south side of Simmond's Hill. These quarries are now enclosed within the landscaped garden of a house known as Manley Knoll and, although extensively planted with rock-garden vegetation, they expose faces of massive building stone up to 5 m high. Higher massive sandstones crop out on the rough ground between Sunnybank Farm and Manley Common, and the highest makes the bold scarred ridge of Black Wood [516 725]. The shelf above Black Wood roughly coincides with the outcrop of the base of the 'Frodsham Beds' facies, and all the high hillocky ground between here and the road west of New Pale Farm is underlain by sandstones of this type.

Among several boreholes into the Helsby Sandstone, those at Newton Hollow Pumping Station [522 735] have already been mentioned. A generalised section of No. 3 Borehole appears in Figure 7, and details are given in Appendix 1, p.104. A borehole [522 735] at the hospital (formerly Liverpool Sanatorium) starts somewhere in the 'Frodsham Beds' facies and pierces strata described as hard red sandstone to a depth of 88 m. Underlying strata described as soft red sandstone are presumably Upper Mottled Sandstone. In contrast, only the top 19 m or so of the Helsby Sandstone seem to have been pierced in a nearby borehole [530 734] at Crossley Hospital (formerly Manchester Sanatorium), which started in strata high in the Tarporley Siltstones. The Geological Survey's Hondslough Borehole [537 726] is also illustrated in Figure 7, and details are given in Appendix 1, p.102. The field relationships of the small outcrops of Helsby Sandstone at Longley Wood and Kelsall village suggest that they lie within 30 m of the base of the Tarporley Siltstones and, therefore, would be expected to be in 'Frodsham Beds' facies. In this part of its outcrop, however, the facies appears to include more massive and well cemented beds. About 7.5 m of massive and cross-bedded yellow sandstone are exposed in a quarry [572 699] in Longley Wood. The lower 3 m are fairly well cemented, but the upper 4.5 m are not coherent enough to have yielded a satisfactory building stone. In a quarry [524 682] on the north side of the A54 in Kelsall village, more than 5 m of well bedded soft yellow sandstone overlie 60 cm of cross-bedded yellow sandstone resting on more than 6 m of fairly coherent, massive yellow sandstone.

Similar yellowish sandstones make the ridge east of, and overlooking, Kelsall, which ends southwards in a bold hill, the site of Kelsborrow Castle. The sandstone in an old quarry [531 684] in Quarry Lane is highly cross-bedded with wedge-shaped masses of hard and soft sandstone. At Kelsborrow Castle 6 to 9 m of massive and cross-bedded sandstone are exposed in crags on the south-west slopes of the hill. In the same fault-bounded outcrop, sandstones of the 'Frodsham Beds' facies make hillocky ground for 100 m or so north of the A54 between two straight lanes that lead up from that road to the Yeld.

The most extensive outcrop of the Helsby Sandstone lies east of Kelsall, between the Primrose Hill Fault and the East Delamere Fault. In this tract the strata lie nearly horzontally, and the basal parts of the Helsby Sandstone crop out only in the Willington area. They were pierced in boreholes at Eddisbury Pumping Station [557 694] and Delamere (Blackbank Wood) Pumping Station [561 677]. Generalised sections of these boreholes are shown in Figure 7, and details are given in Appendix 1, pp.99, 101. Their correlation is discussed on p.26.

The middle part of the Helsby Sandstone is exposed in numerous natural scars and in several old quarries. Prominent natural scars occur between Organsdale House and King's Chair (Figure 10), and around the north and east sides of Primrose Hill, in dry gorges known as Urchin's Kitchen [541 679] and Dog Fall [547 678]. Very bare terrain with numerous exposures lies south of Delamere School and extends south-eastwards for 1 km or so between the A54 and Hindswell Gutter. Almost all these natural exposures show hard, massive or cross-bedded, reddish brown sandstone with scattered quartz pebbles, and several contain lenticular beds of clay-gall conglomerate. Eddisbury Quarry [558 693], the main face of which extends for nearly 200 m both to the north and south of Stoney Lane, exposes 9 m of hard, massive and cross-bedded brown sandstone with many quartz pebbles. Quarries [561 685] to the east and south-east of St Peter's Church, Delamere, expose 6 m of more of similar sandstone. North-west of Delamere Rectory (Figure 10) the following strata are visible [566 687] in the largest of a number of old excavations:

|  | Thickness m |
|---|---|
| Massive reddish brown sandstone with quartz pebbles | 4.50 |
| Flaggy red sandstone with bands of red sandy shale | 1.22 |
| Red, shaly sandstone with thin grey bands | 0.61 |
| Wedge-bedded, red sandstone with partings of red mudstone | 1.80 |

The highest parts of the Helsby Sandstone, all of 'Frodsham Beds' facies, are very well exposed in characteristically featured terrain surrounding the outlying mass of Tarporley Siltstones which forms Pale Heights and Eddisbury Hill (Figure 10). At the north-west corner of the outlier, about 500 m north of the water tower on Pale Heights, strata high in the 'Frodsham Beds' facies are exposed in crags [543 701] that reach a height of 6 m or so. The same strata make the steep bank below the western crest of Pale Heights which merges southwards and south-eastwards into a less conspicuous feature, below and to the south of which lies a broad expanse of hillocky country formed by the lower part of the 'Frodsham Beds' facies. The top of the 'Frodsham Beds' facies is visible [548 690] on the north side of a dry channel occupied by a farm track 400 m south-west of Old Pale Farm. Here 6 m or so of soft, reddish brown, highly cross-bedded, sandstones are sharply overlain by soft red sandy mudstones, the basal Tarporley Siltstones.

The bold eastern [555 694] and southern [553 693] slopes of Eddisbury Hill expose more than 15 m of strata, virtually the whole of the upper part of the 'Frodsham Beds' facies. The many small crags on the steep hillside are composed of fairly soft, reddish brown, highly cross-bedded sandstone, usually containing numerous small, harder nodules cemented with barite. The top of the Helsby Sandstone lies very close above the highest strata exposed in a cliff [553 693], about 6 m in height, that makes the southern rampart of the hill-fort.                                                                    JRE

## The central area (Willington to Eaton)

The bold west-facing escarpment of Birch Hill and Willington Wood, made by the middle part of the Helsby Sandstone, is transected by a deep ravine east of Willington Corner. The road from Willington Corner to Roughlow Farm ascends high on the north bank of the ravine, and provides the best section in this area. The section is continued eastwards in the upper part of the ravine. The roadside section begins where the road swings up into the ravine [535 670] and good exposures on its north side extend up to an elbow bend [536 671] where the road swings out of the ravine to run northwards towards Roughlow Farm. Something approaching 30 m of the middle parts of the Helsby Sandstone, including pebble beds and clay-gall conglomerates, are exposed dipping very gently eastwards. The section continues in a number of exposures of conglomeratic sandstone which form a crag [537 671] in the ravine about 45 m east of the elbow bend in the road. These beds are overlain by flaggy sandstone and sandy mudstone, similar in lithology to Tarporley Siltstones although well down in the Helsby Sandstone, which can be seen [538 671] near a small reservoir at the head of the ravine, about 140 m east of the elbow bend.

South of Willington Corner, near the entrance to Willington Hall, there is another, but much shorter section in the road [535 662], and in an old quarry just west of Park Farm. These sandstones are probably on about the same horizon as the lowest sandstones exposed in the section to the north.

Higher in the sequence, sandstones of 'Frodsham Beds' facies make gentle mound-and-hollow features below the escarpment east of Tirley Garth [545 663]; the steep, 'tumbled', south and west slopes of Tirley Hollow, below Tirley Farm [549 667], are made by the same strata. Farther east, a broad outcrop of 'Frodsham Beds' facies makes distinctive topography around Forest Farm.

A deep ravine known as Holbitch Slack exposes [566 658] a fault about 100 m east of the road that crosses the head (western end) of the ravine. The fault throws down Tarporley Siltstones on its west side against Helsby Sandstone of 'Frodsham Beds' facies to the east. Some 21 m of these highly cross-bedded sandstones are well exposed for over 300 m along the ravine to the east of the fault. They are nearly horizontal and, in one place, Strahan (1882, figure 2) saw an exposure of the top of the 'Frodsham Beds' facies. JRE, DM

Generalised sections of the Helsby Sandstone proved in a borehole at Cotebrook and in another at Eaton Pumping Station are shown in Figure 7; details are given in Appendix 1, pp.98, 100. Borehole No. 2 at Eaton is probably the best recorded section of the Helsby Sandstone in the whole district. At the base a fine-grained sandstone with ripple-marked, sun-cracked and bioturbated muddy partings is overlain by reddish brown sandstone with clay galls. Numerous clay-gall conglomerates occur from 13.5 to 33.5 m above the base and pebbly sandstones are common from 21.5 to 70 m above the base. A fine-grained sandstone lying 68.5 m above the base is recorded as having 'an irregularly eroded and potholed upper surface', and a bed of grey sandstone, 74.5 m above the base, is noteworthy in that it yielded impressions of plant stems up to 61 cm in length. Sandstones of 'Frodsham Beds' facies comprise the uppermost 23 m of the formation, the top of which is marked by the usual abrupt change of lithofacies. Some of the sandstone in the middle part of the 'Frodsham Beds' facies is recorded as containing concretions of barite.                                                          JRE

## The south (Beeston to Bolesworth)

Overlying the Upper Mottled Sandstone, 15 m of red-brown, well cemented building stones with subangular grains, red-brown shaly partings, and a few softer bands of brighter red 'millet-seed' sandstones are well displayed north of Beeston Castle Grotto [538 590]. These are the so-called 'Passage Beds' from the Upper Mottled Sandstone into the Helsby Sandstone, which make a good feature that can be traced round Beeston Castle Hill to a roadway gully

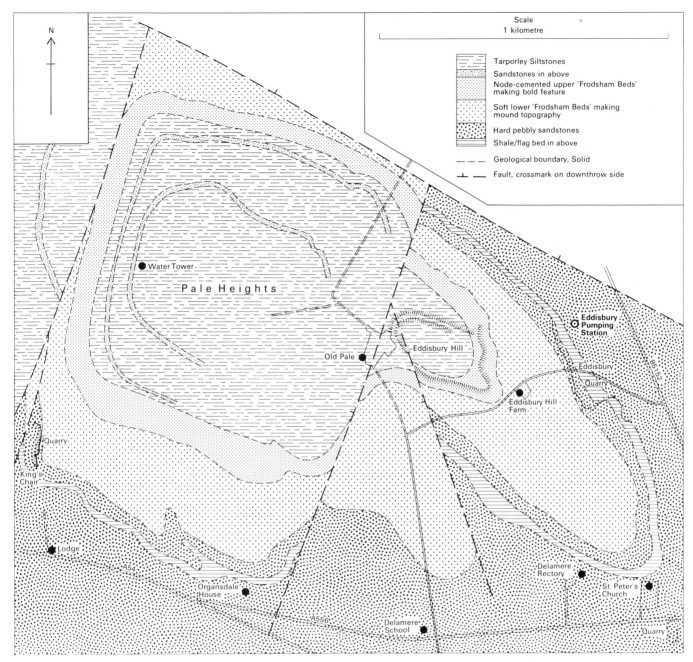

**Figure 10**   Map of the outcrops of the 'Frodsham Beds' facies and middle part of the Helsby Sandstone around the Tarporley Siltstones outlier of Pale Heights and Eddisbury Hill, Delamere

45 m north-west of the Castle Lodge. Here 30 cm of basal, hard, red-brown sandstone rests on 2.5 m of soft, red Upper Mottled Sandstone. These sandstones are rich in barite and have undergone gentle folding and slickensiding due to the proximity of the Peckforton Fault. The dip of the Helsby Sandstone increases from 5° to E8°S at the summit of the hill, to 12° at E27°S in the roadway gully exposure north-west of the Lodge. The building stones make prominent bluffs upon which the curtain walls of the outer fortification of Beeston Castle are built (see Figure 9). The overlying Sandstone Conglomerate forms the cap rock upon which the keep and castle [537 593] are built. The rock is grey to grey-pink in colour, markedly cross-bedded, very coarse-grained and pebbly. There are good exposures of the Sandstone Conglomerate on the south and east sides of the castle and in the moat; the north and west sides are near-vertical, inaccessible crags. Altogether, up to 9 m of the Sand-

stone Conglomerate are seen under and around the keep and castle; the dip is E8°S at 5°.

Good exposures of the 'Passage Beds' and Sandstone Conglomerate are seen on the bold escarpment north and west of Peckforton Castle [533 580]. Some 145 m north of Peckforton Castle, red-brown Helsby Sandstone is seen in a path leading to Horsley Lane, the Sandstone Conglomerate making a prominent feature on the south side of the path. Pebbly sandstone of this unit is seen in a gully 45 m north of the north-west corner of the castle. The castle (see Figure 9) is built on the red-brown, well cemented, coarse, angular-grained cross-bedded sandstone, apparently without pebbles, which forms the upper part of the Sandstone Conglomerate sequence. Up to 3 m of this type of sandstone are seen dipping E8°N at 7° in the eastern footings of the castle walls. In the moat on the west side of the castle, about 2 m of similar sandstone dips at 7° in the same direction. Nearly 385 m east of Peckforton Castle, an old quarry [537 581] just west of the Beeston–Peckforton road, exposes about 6 m of fine- to medium-grained, red-brown sandstone dipping E10°N at 7°. This sandstone is probably only slightly higher in the sequence than those on which the castle is built, because of the considerable down-dip fall in ground level. The lower part of the sandstone is massive-bedded, but the upper part is more thinly bedded and fissile.

At The Table Rock [530 580], 200 m south-west of Peckforton Castle, 3 m of red-brown, coarse-grained, well cemented basal Helsby Sandstone are overlain by 7.5 m of massive and erosion-resistant, red-brown, cross-bedded, pebbly Sandstone Conglomerate. The junction between the Helsby Sandstone and the Upper Mottled Sandstone is obscured by downwash along this part of the scarp. South-westwards along the scarp from The Table Rock, continuous exposures of up to 3 m of conglomeratic grit are seen for a distance of 275 m. Some 180 m north-west of Stanner Nab [530 575], about 3 m of shaly, red-brown, cross-bedded 'Passage Beds' rest on 3 m of massive, cross-bedded, red-brown Upper Mottled Sandstone with a high proportion of coarse 'millet-seed' grains. The Sandstone Conglomerate at Stanner Nab forms a prominent scarp on the upthrown south side of an E–W fault (see Figure 9), and the scarp exposes 10.5 m of hard, red-brown, coarse-grained, well jointed, cross-bedded, pebbly sandstone. The pebbles are scattered in the upper part and become more abundant downwards, but the junction with the underlying 'Passage Beds' is not exposed.

Along the west-facing scarp of the Peckforton Hills, to the south of Stanner Nab, the lower pebbly sandstones of the Sandstone Conglomerate are well exposed for a distance of 290 m, and about 9 m of red-brown sandstone with scattered pebbles are seen in an old quarry [530 572] at the southern end of these exposures. Good exposures of the uppermost beds of the Sandstone Conglomerate, with rare pebbles, are seen in old quarries [532 575] west of Stanner Cottages. These expose very hard and erosion-resistant, massive and cross-bedded, well jointed, red-brown building stones. The hardness may be due to mineralisation associated with the Stanner Nab Fault, for nearby the sandstones immediately adjacent to the fault are converted to quartzite, probably by thermal and dynamic metamorphism, and altered from their normal dark, red-brown colour to a light grey-brown.

On the main range of the Peckforton Hills south of Stanner Nab, scattered exposures of the uppermost non-pebbly beds of the Sandstone Conglomerate are common. The best of these exposures is in an old quarry [530 563], 455 m north-east of the Lodge at the head of Peckforton Gap (see Figure 9). In this quarry up to 3 m of massively bedded, well jointed sandstone, overlain by 1 m of flaggy sandstone dip a little west of north at 5°. An old well situated within the quarry still yields water in spite of the heavy pumping at the Peckforton Gap boreholes situated some 900 m to the south. This suggests the presence of a perched water table within the Helsby Sandstone. The sandstone exposed in a 3.5 m cutting at the entrance to the quarry is well bedded and jointed, but presents a rather rubbly appearance in places.

Along the main east-facing scarp of the Peckforton Hills southwest of Peckforton village, from 3.5 to 6 m of coarse conglomeratic sandstone are continuously exposed, and at Peckforton Point [530 558] some 15 m of the basal 'Passage Beds', dipping NW at 12° to 15°, underlie 6 m of coarse conglomerate. The 'Passage Beds' are brown and coarse-grained, hard and well cemented, but with intervening shaly bands. They are markedly cross-bedded and rest conformably upon softer, finer-grained, red and fawn Upper Mottled Sandstone. The overlying conglomerate has a sharp, irregular base, and incorporates fragments of the underlying rocks. It is hard and well-banded generally, brown in colour, but with a white baritic cement — though in some places the cement is of red iron

oxide — and it is all markedly cross-bedded. The matrix is predominantly composed of subangular coarse brown grains with a few 'millet-seed' grains, and the pebbles are well rounded, light grey and 'liver-coloured' quartzites up to 5 cm across.

The Helsby Sandstone, including the Sandstone Conglomerate, is well exposed in a roadway by the Lodge [526 559] at the head of Peckforton Gap. In an old quarry 45 m south of the Lodge, 4.5 m of massive, non-pebbly, hard brown basal Helsby Sandstone are seen, and up to 4.5 m of flaggy sandstone with mudstone bands, resembling beds of the Mercia Mudstone Group overlie the Upper Mottled Sandstone on the south-western bank of Peckforton Gap. The mudstone bands probably lie just below the sandstone exposed in the quarry.

An isolated exposure of brown pebbly Helsby Sandstone is seen [542 568] some 275 m north-west of Peckforton Hall. This sandstone is 3.5 m thick and dips NW at 15° but is badly jointed and broken.

In the Burwardsley area the Sandstone Conglomerate forms the cap rock of the west-facing scarps. The following section is seen [519 566] at the north end of Willow Hill (see Figure 9):

| | *Thickness* m |
|---|---|
| SANDSTONE CONGLOMERATE | |
| Coarse brown sandstone with a few small scattered pebbles | 2.0 |
| Gap | 0.5 |
| Massive and markedly cross-bedded, coarsely conglomeratic brown grit with quartzite pebbles up to 5 cm across | 7.5 |
| 'PASSAGE BEDS' | |
| Soft, fine-grained, shaly brick-red sandstone with fawn mottling | 1.0 |
| Hard, medium-grained angularly weathering brown sandstone | 1.0 |
| Gap | 3.5 |
| Hard, shaly, red-brown sandstone, angularly weathering, with a few bands of coarse, 'millet-seed' sandstone, and micaceous layers | 3.0 |
| Gap | 1.5 |
| UPPER MOTTLED SANDSTONE | |
| Soft bright red and fawn 'millet-seed' sandstone | 12.0 |

These beds dip due east at about 5°, and good exposures of Sandstone Conglomerate are again seen [518 564] in the south-western side of Willow Hill where some 6 m are exposed. Nearby, in an old quarry 135 m south-east of Quarry House [518 562], some 4.5 m of hard, well jointed, massive bedded brown sandstone dip E30°N at 7°.

The west-facing scarp of Burwardsley Hill provides many sections in the lower part of the Helsby Sandstone. In the north-east corner [511 561] of Cawley's Wood, about 2.5 m of coarse, brown and grey sandstone with a few scattered pebbles form the topmost crags and dip NE at 7 to 10°. About 6 m of section below this sandstone are obscured by downwash; then 2 m of soft, coarse-grained, red, brown and grey conglomerate with abundant white quartzite pebbles, the basal part of the Sandstone Conglomerate rests with an erosive base on 2 m of hard, light fawn, fairly coarse-grained, well cemented sandstone, belonging to the 'Passage Beds'. These sandstones dip north-eastwards at 10°, but 55 m along the scarp to the south-west the beds are gently folded into a broad anticlinal structure, the west limb of which dips south-westwards at 7°. Tension jointing at right angles to the dips is well developed on the crest of this shallow anticline, where the following section [510 561] is seen:

|  | *Thickness* |
|  | m |

SANDSTONE CONGLOMERATE

Sandstone, massive, coarse-grained, conglomeratic
  with a few lenses of very hard, non-pebbly sandstone,
  markedly cross-bedded, grey and leached, base
  sharply erosive — 7.62

'PASSAGE BEDS'

Shale, brick red with a few green mottles, sandy,
  coarsely micaceous — 0.61
Sandstone, brick red, fawn and brown, massively
  bedded, coarse-grained with abundant 'millet-seed'
  grains — 2.44
Shale, brick red, sandy, micaceous — 0.30
Sandstone, brick red, micaceous, medium-grained
  with thin bands of sandy shale — 0.46
Sandstone, hard, brown, coarse- and angular-grained,
  compact; base not seen — 0.46

A few yards farther along the scarp to the south-west the measures are again gently folded so that the dip resumes its north-easterly direction at 5 to 10°. At the crags above Sandhole Farm, [509 560], 4.5 m of coarse, dark brown to light grey, markedly cross-bedded Sandstone Conglomerate overlie: red sandy mudstone 61 cm; on brown, coarse, 'millet-seed'-grained, cross-bedded sandstone, the top metre rather resembling Upper Mottled Sandstone, 4.5 m; on fawn, fairly fine-grained, shaly sandstone 0.9 m; over hard, brown, well cemented building stone, of which 1.2 m is exposed. The junction of these 'Passage Beds' with the underlying Upper Mottled Sandstone probably just underlies the lowest exposure and is obscured by boulder clay. On this scarp the best exposures of Sandstone Conglomerate are seen in crags [508 559] above Hillside Farm where about 7.5 m of coarse, grey, pebbly grit has pebbles, (generally under 2.5 cm across) of green and brown quartzite and white quartz, disposed in strings and lenses which emphasise the marked cross-bedding. Beneath the pebbly grit, the following 8 m-section of 'Passage Beds' showing only minor variation from the previous section is, in ascending sequence: red, sandy mudstone 15 cm; unexposed 76 cm; hard, dark grey-brown, coarse-grained sandstone 2.5 m; softer, poorly exposed, red-brown sandstone of 'mottled sandstone' type 1.5 m; hard, cross-bedded, red-brown building stone 2 m; bright red, fine-grained micaceous sandstone 1.2 m. These beds rest upon Upper Mottled Sandstone of which about 1.5 m are exposed.

Farther along the scarp of Burwardsley Hill to the south-west, up to 7.5 m of Sandstone Conglomerate are continuously exposed, the dip diminishing until the beds are nearly horizontal at the triangulation point [508 557] 430 m WSW of Burwardsley Hill Farm. South of the triangulation point, the crags above Mickerdale Farm expose up to 12 m of massively bedded, coarse-grained, pink and grey grit with many pebbles in the lower 7.5 m. The grit is markedly cross-bedded with a few lenses and bands of finer-grained sandstone set among coarse, pebbly grit bands. On the crags [508 557] 135 m south-east of Mickerdale Farm, a 5 cm sandstone dyke is intruded into a bed of massive, coarsely conglomeratic, highly hematitised grit, apparently along a joint plane. South-eastwards along the scarp from these exposures, up to 9 m of Sandstone Conglomerate are continuously exposed for a further 90 m and there are small exposures of underlying 'Passage Beds' in a few places.

Burwardsley Hill Farm stands on coarse brown sandstone, and a small stream to the east and south-east of the farm exposes part of the Sandstone Conglomerate and some underlying sandstones. In the eastern bank [516 559] of the stream, 320 m due east of Burwardsley Hill Farm, 1.5 m of brown rather flaggy, coarse-grained sandstone with some 'millet-seed' grains rest on 3 m of coarse brown grit. This grit is markedly cross-bedded in places and contains small quartz pebbles up to 1 cm across, though one pebble of 4 cm in diameter was noted. The exposures probably lie about 9 to 12 m above the base of the Sandstone Conglomerate; they extend for about 65 m downstream, and the dip is NE to NNE at 5°. The same strata are seen at a small waterfall in a tributary about 18 m east of the main stream and some 32 m above the confluence; they are also seen in another tributary about 45 m to the north, near a bend in the main stream. These exposures stratigraphically overlie upstream exposures [515 557] of shaly sandstone and non-pebbly brown sandstone, about 270 m south-east of Burwardsley Hill Farm, which probably belong to the 'Passage Beds'.

On the scarp east of Bolesworth Castle [496 560] some 9 m of coarse, brown, cross-bedded grit rest directly upon 2.5 m of typical Upper Mottled Sandstone, there being no intervening 'Passage Beds'. More exposures of Helsby Sandstone are seen in an adjacent road cutting [495 558]. In all it is estimated that no fewer than 24 m of Helsby Sandstone are exposed near Bolesworth Castle, the lower 18 m containing pebbles. About 38 m higher in the sequence, a stream section [499 557] 230 m east of Bolesworth Hill Farm shows 6 m of well jointed, massively bedded, red, 'millet-seed' sandstone closely resembling Upper Mottled Sandstone in appearance. Still higher in the sequence, 365 m downstream, 2 m of rather flaggy red sandstone are exposed [500 560]. These are the highest exposures of Helsby Sandstone in the southern part of the district and are probably of 'Frodsham Beds' facies. The top of the formation, beneath the Tarporley Siltstones of Harthill (Poole and Whiteman, 1966, p. 34), lies immediately to the south-east. EGP

CHAPTER 4

# Triassic: Mercia Mudstone Group

The current classification of this part of the Triassic sequence in the Chester district is outlined in Figure 11.

Theories about the origin of the Mercia Mudstone Group (formerly 'Keuper' Waterstones and 'Keuper' Marl) have developed greatly over the last two decades and are still evolving. It is, therefore, inappropriate at this stage to attempt a definitive account of the depositional history and environment; nor is this the place in which to argue the merits of conflicting views. It is sufficient to note them where they are relevant. Since the previous sheet-memoirs dealing with parts of the Cheshire–Shropshire basin were written there have been two most important advances. These are, first, an increased understanding of major crustal processes and their effect on post-Carboniferous sedimentation in this region, and second, the application of micropalaeontology, particularly palynology, to the study of British Triassic sequences that were long regarded as unfossiliferous. The first has provided an environmental and tectonic framework in which to place the known sequences and structures, and the second, particularly through the work of G. Warrington (1967 and the following), has inserted time planes into what was previously a purely lithostratigraphical compilation, allowing a series of relistic palaeogeographies to be constructed for various periods in the Trias.

## SEDIMENTARY ENVIRONMENT AND LITHOFACIES

During the deposition of the Sherwood Sandstone Group the Worcester and Cheshire grabens formed a continuous valley that continued to subside as it was supplied with sediment by a major river system originating in the uplands of northern France. As this period of deposition was ending the Chester district lay at an approximate palaeolatitude of 19° north, in an arid climatic zone in the heart of a great continent, far from the sea. The rift system was one of several 'failed arms' in the tectonic activity that culminated in the opening of the north Atlantic (see Structure, p. 57).

The base of the Mercia Mudstone Group (see Figure 11) records a sharp and profound change in sedimentation from the fluvial and aeolian deposits of the Sherwood Sandstone Group to a regime strongly influenced by periodic influxes of sea water. The Tarporley Siltstones (Waterstones) represent a widespread marine transgression over the Sherwood Sandstones. The view expressed by Audley-Charles (1970, p. 60) that this transgression equates in time with the deposition of the Muschelkalk in the European continent and the North Sea seems, however, incorrect, for Warrington (1970b, p. 212) considered that the southward progress of the 'Waterstones' facies may have taken a long time – from the late Scythian in Cheshire to the late Anisian in Worcester-

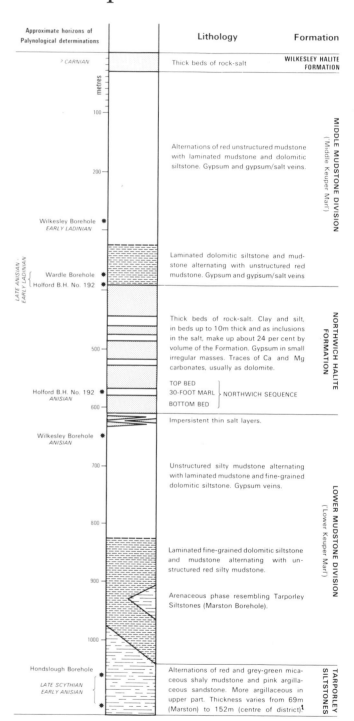

**Figure 11**  Generalised vertical section of the Mercia Mudstone Group in the Chester District

shire – though in this latter area the facies is rather different. The transgression in the Cheshire and west Midlands area may have been assisted and prolonged by progressive crustal sagging along graben axes.

The Tarporley Siltstones are thinly-bedded micaceous sandstones, siltstones and silty mudstones, with mica concentrated on some bedding planes and much argillaceous material, both in mudstone layers and incorporated in the sandstones and siltstones. There is strong evidence that this facies in north Cheshire is intertidal: the detailed description of the tidal processes which gave rise to a small area of 'Waterstones' facies near Daresbury, by Ireland and others (1978, pp.399 – 436), applies equally well to the Tarporley Siltstones of the Chester district. These authors envisaged a tidal flat environment varying from low intertidal sandflats to high intertidal mudflats, with sand bars in tidal flat channels. The sediment surface frequently dried out, particularly in the high tidal zone, producing mudcracks, mudflakes and crystals of halite whose form is preserved in pseudomorphs of mud or silt.

Unlike the Sherwood Sandstone Group these sediments appear to owe little to distant southerly sources; the river system which transported the earlier coarser sediments northwards towards the Cheshire basin and beyond may have been losing its influence because of peneplanation of its upper reaches. The Mercia Mudstone Group is largely the product of the lateritic weathering, erosion and peneplanation of the surrounding arid areas of the Welsh and Pennine uplands. Fitch and others (1966, p.309), on the basis of the isotopic ages of detrital micas, concluded that whereas the 'Bunter' (i.e. Sherwood Sandstone) micas of north-east Cheshire were derived via the Worcester Graben from the far south, those in the 'Keuper' (i.e. Mercia Mudstone) were of mixed derivation, some from the earlier 'Bunter' sources and others from more local sources, including the reworking of earlier Triassic, Permian and Carboniferous sediments. Some sediment was presumably carried in by streams, but wind was an increasingly important factor in the transport of the material of dust grade which constitutes much of the Group. The dominantly red colour of the dust particles is due to an iron oxide pellicle apparently finely disseminated over the surface of individual grains. Where the rock is grey or grey-green either from primary or secondary causes the pellicle has been chemically removed, in the one case before, and the other case after, deposition.

The upward passage from the Tarporley Siltstones into the overlying mudstones ('Keuper Marl') reflects a gradual change from deposition almost wholly in water to a regime in which aqueous sedimentation alternated with deposition on dry or almost dry ground. At the same time the proportion of material carried into the basin by wind increased in comparison to that transported by water. Descriptions of the lithological features of the 'Keuper Marl' in memoirs of the Stockport (98) and Macclesfield (110) districts apply equally to the lower mudstone division in this district. Generally speaking, the same sedimentary types are present in the mudstone divisions as in the Tarporley Siltstones below, but with less arenaceous and more argillaceous material. An irregular rhythm is discernable in the stratification; each

sedimentary cycle, where well developed, has laminated strata or 'striped beds' at the base, with fine-grained and commonly dolomitic siltstones in thin well sorted and sharply defined layers alternating with mudstone, which may be of exceedingly fine and even grain. Locally, the banded strata are found to consist of alternations of light and dark mudstone, the paler layers being somewhat coarser in grain than the darker, and it is in the banded strata that such features as mudcracks and the pseudomorphs of halite crystals are chiefly seen. The banded strata, which may be either red or grey, are overlain in each cyclic unit by 'blocky' or unstratified mudstone which appears to owe its lack of structure to the fact that it has been broken up, mixed and allowed to settle before final consolidation took place. This alternation is repeated many times throughout the Mercia Mudstone sequence above the Tarporley Siltstones, being interrupted only by the presence of thick halite layers.

The cyclic nature of the Mercia Mudstone sedimentation has been examined in detail by Arthurton (1980, pp.43 – 58) with the aid of sequences from three boreholes, one in this district near Winsford and two south of Middlewich in the Macclesfield (110) district to the east. He found that desiccation cracks are most abundant in the upper parts of laminated units, and tend to occur directly under units of the blocky facies, material of which plugs the fissures. Arthurton considered that each change from laminated to blocky lithology reflects a change from subaqueous to emergent conditions, and supported Taylor's (1963, p.74) conclusion that the banded strata represent material carried in dust storms and blown into water, whereas the blocky unstratified rock is the result of dust deposition on dry or nearly dry ground, as in the 'ploughed ground' or 'zardeh' which forms extensive flats around contemporary salt lakes in central Iran, and in which the growth of evaporite minerals breaks up the surface of the ground, destroying any previous sedimentary structure.

Explanations for the rhythmic alternations of facies have ranged from a jerky or pulsating basin subsidence to variations in climate. The sedimentary regime appears in some ways analogous to the cyclic sedimentation of the Coal Measures (Westphalian) in which a deltaic area was repeatedly submerged after subsidence and built up to sea level again by deposition. A favoured explanation in that case was steady subsidence of the deposition area, coupled with eustatic rises and falls of sea level. A similar explanation might well hold good in the case of the Mercia Mudstones.

There has been some discussion as to whether or not halite beds represent total desiccation and emergence. Wills (1970, p.280) suggested that the evaporite-rich sequences may record frequent or prolonged exclusion of the sea from a shelf that is normally flooded by sea water, but he did not imply that there needed to be evaporation to dryness for the production of salt. Evans (1970, p.114) believed that in the evaporitic sequences mudstone and halite alike were subaqueously deposited from a bottom layer of saturated brine, each salt bed representing a short period when temporarily supersaturated bottom brine was brought back to saturation by the precipitation of salt. Arthurton (1980, p.53) found signs of both subaqueous and subaerial deposition amongst the mudstones in evaporitic as well as non-evaporitic sequences. In particular he cited the existence of layers of structureless blocky mudstone — interpreted as loess accumulated on almost dry ground — as clear evidence of emergence. These conflicting views have yet to be resolved; there is no need to discuss them further so far as this account is concerned. Whatever explanation is eventually accepted, it must take account of the fact that thick halite beds were laid down, probably quite rapidly, in such a uniform manner that they can be correlated, together with their separating mudstone layers, almost bed for bed from one end of the basin to the other.

Although sedimentation of the Mercia Mudstones probably took place over a very wide shelf area, filling all the ground between the Lower Palaeozoic upland massif of Wales on the west and what are now the central uplands of England on the east, erosion has removed much of the evidence away from the deepest fault troughs where the thickest sequences accumulated. That part of the deposition area which lay between the East Delamere Fault and the Red Rock Fault (the Cheshire Graben of White, 1949) was one of these most active areas. The faults bounding this trough were operating from at least early in the period of Mercia Mudstone deposition. Sedimentation kept pace with subsidence and at least 2000 m of sediments and evaporites of this Group accumulated in the deepest parts of the trough. Evans (1970, p.110) found evidence, in certain variations in the saliferous sequences near the edges of the saltfield, that the present structural margins of the fault trough—the Red Rock Fault on the east and the East Delamere Fault on the west—lie close to the original depositional limits of the salt.

## PALYNOLOGY

Grey-green mudstones from the Tarporley Siltstones in the Hondslough Borehole were examined for palynomorphs and assemblages of miospores comparable with those known from the upper Bunter (Röt) and lower Muschelkalk in Holland and Germany were recovered from 20.19 and 65.53 m. The latter includes *Densoisporites nejburgii* and *Retitriletes jenensis* and is comparable with assemblages from the Hambleton Mudstone Formation at the base of the Mercia Mudstone in west Lancashire (Warrington, 1974) and may indicate a correlation with that formation; it is assigned a late Scythian to early Anisian age (Figure 11). The assemblage from 20.19 m includes *Stellapollenites thiergartii*, *Illinites chitonoides* and *?Angustisulcites gorpii*; it is comparable with those known from the Singleton Mudstone Formation in west Lancashire and is assigned an Anisian age.

Samples from the 'Waterstones' in the Wilkesley Borehole in the Nantwich (122) district proved barren but an assemblage of miospores including *Perotrilites minor* and *Stellapollenites thiergartii*, indicative of an Anisian age, was recovered from the 'Waterstones' at Liverpool (Fisher, 1972a, b).

No palynomorphs have been recovered from the lower mudstone division in the district. In the Nantwich (122) district samples from 17.68 m above the base of the unit at 1509.83 m in the Wilkesley Borehole yielded a few miospores indicative only of late Scythian to Anisian age (Warrington, 1970a). An assemblage from 1306.98 m, 37.19 m below the top of the unit in the same borehole (Figure 11) is dominated by miospores, principally *Alisporites grauvogeli*, but includes a few specimens of an acanthomorph acritarch, *Micrhystridium*, which are indicative of an aqueous depositional environment subject to marine influence; the assemblage is assigned an Anisian age.

No palynomorphs have been recovered from the Northwich Halite in the district but three small assemblages of miospores regarded as Anisian in age were recovered from the '30-Foot' Marl in ICI Borehole 192 at Holford in the Macclesfield (110) district (Warrington, 1970a, b).

Samples from the middle mudstone division at 86.26 m in the Worleston Borehole and 147.52 m in the Wardle Borehole proved barren. An assemblage from the basal part of this unit at 183.19 m in the latter borehole yielded a very poorly preserved assemblage of miospores including *Tsugaepollenites oriens* which is indicative of an Anisian to early Ladinian age and of a correlation with horizons both below and above the Preesall Halite in west Lancashire (Warrington, 1974). An assemblage recovered from the middle mudstone division at 101.19 m in the Over Borehole comprises a few indeterminate bisaccate miospores and relatively numerous Tasmanaceaen algae; it affords no evidence of age but the algae are indicative of deposition in an aqueous environment subject to marine influence.

A sample from the basal bed of the middle mudstone division in an ICI Borehole at Holford in the Macclesfield (110) district yielded a more varied miospore assemblage than that recovered from slightly higher beds in the Wardle Borehole; the Holford assemblage includes *Tsugaepollenites oriens* in association with *Perotrilites minor* and is assigned a late Anisian age. Samples from 967.74 m, 109.73 m above the base of the unit, in the Wilkesley Borehole in the Nantwich (122) district, yielded miospore assemblages, including *T. oriens* in association with *Protodiploxypinus potoniei*, which are regarded as Ladinian, possible early Ladinian, in age. Sporadic specimens of a polygonomorph acritarch, *Veryhachium*, are present in these assemblages and are indicative of deposition in an aqueous environment subject to marine influence.

No palynomorphs have been recovered from the higher part of the middle mudstone division or from representatives of any higher units in the Mercia Mudstone succession present in the district and in neighbouring areas.

## TARPORLEY SILTSTONE

The lower 60 to 90 m of the Tarporley Siltstones consist of alternations of dark red or secondarily green, micaceous shale, and pink, fine-grained, argillaceous, well sorted, micaceous, ripple-bedded, ripple-marked sandstone. Mudcracks are common and some bedding planes show worm tracks and burrows (Pollard, 1981), and footprints of *Chirotherium* and *Rhynchosauroides*. Other features include hopper pseudomorphs after halite, pinhole cavities (?after gypsum) and rare calcareous nodules. The major mudstone/sandstone alternations, which give repeated contacts of pervious and impervious layers, are responsible for the occurrence of seepages of water, in places copious, wherever the strata are exposed in down-dip cuttings or quarries. This is a possible origin of the name 'Waterstones' which was previously applied to this formation (Strahan, 1882, p.9). The hard and soft layering also gives rise to ridge-and-slack topography in drift-free areas and, where the strata are nearly horizontal, a 'table-land' type of landscape results.

Higher in the sequence the beds of sandstone are thinner, and there is a very gradual passage to the mainly red, silty mudstones with thin silty sandstone (skerry) bands of the lower part of the lower mudstone division of the Mercia Mudstone Group. The boundary between the Tarporley Siltstones and this lower mudstone division is thus an arbitrary one, and there is no general agreement as to where it should be drawn. The Ravenslodge Trial Borehole for Newton Hollow Pumping Station reached the base of the Tarporley Siltstones at 132 m. The cores were examined by W. B. Wright, D. A. Wray and R. C. B. Jones in 1937 and they allocated 70 m of strata to the 'Waterstones'; the higher beds were referred to the 'Keuper Marl'. At variance with their conclusions is the fact that strata at or close above the horizon at which the borehole began are exposed in marl pits on Finney Hill, and contain at least one thick bed of sandstone which makes a bold feature around the adjacent hillside. This is contributary evidence for the inclusion of the strata within the Tarporley Siltstones; indeed the marl pits are shown on the Old Series geological map to occupy a position low in the 'Waterstones'. A nearby borehole at Crossley Hospital (formerly Sanatorium) proved the base of the Tarporley Siltstones at 134 m. If the driller's description of the strata is reliable all the beds down to this depth are of 'Waterstones' facies. The reliability of the log is supported by various surface exposures south of the hospital, all of which include enough sandstone to be classed as Tarporley Siltstones. Thus the thickness of the formation in this area is more likely to be of the order of 152 m than the much smaller figure given by Wright, Wray and Jones; indeed Jones states that, in the Runcorn (97) district to the north, bands of sandstone similar to the 'Waterstones', and distinct from 'skerry' bands, are present above the base of the 'Keuper Marl' (Jones in Wright, 1938, p.44, see also Simmonds in Wedd and others, 1923, pp.91–92). This thickness is substantially more than that recorded at the Marston Borehole (68 m), just outside the north-eastern extremity of the district, and raises the question of whether the Tarporley Siltstones of the Chester district include all the strata up to the top of the 'Upper Keuper Sandstone', a thickness which amounts to 220 m at Marston. Because of the arbitrary definition of the top of the Tarporley Siltstones this matter has not yet been resolved.

Much farther south, the bold hilly country west of Eaton, the largest drift-free outcrop of Tarporley Siltstones in the district, provides evidence for the minimum thickness of the formation. The strata are near-horizontal, and the hill-tops must preserve measures that are at least 50 m higher in the sequence, and possibly more, than the 35 m of 'Waterstones' proved in boreholes at Eaton Pumping Station. This gives a minimum thickness of 85 m, disregarding minor downwarps that may accommodate much greater thicknesses.

Little is known of the Tarporley Siltstones at outcrop south of Tiverton where they disappear beneath thick drift. Some 5 m or so of strata referred to this group crop out near Castlegate Farm, Peckforton.

## Details

*The north: (Alvanley to Delamere)*

There are little more than small isolated exposures of a metre or so of flaggy sandstone in the Tarporley Siltstones outcrop of the Helsby-Manley outlier, as for instance in a brook [493 735] about 450 m north of Abbot's Clough and on a hillside [491 740] about 600 m west of St John's Church, Alvanley. Some 2 m of flaggy sandstone overlie 1 m of red and green marl in an old pit [495 748] about 750 m NNW of the church. Larger marl pits higher up the dip-slope of Helsby Hill are outside the boundary of the district.

·At the north end of the main outcrop, sections proved in Ravenslodge Trial Borehole [531 742] and a borehole [530 734] at Crossley Hospital (formerly Sanatorium) are given in Appendix I, p.99. There are many exposures in lanes and old marl pits west of Castlehill House [538 745], all of chocolate-coloured flaggy sandstones with bands of marl and, at the east end of Finney Hill, a large marl pit [536 741] shows:

|  | Thickness m |
|---|---|
| Coarse-grained reddish brown thickly bedded sandstone | 2.0 |
| Greenish grey and chocolate marl-shale with two thin sandy beds near the base | 4.5 |
| Chocolate flaggy sandstones with marly layers | 2.5 |

This is probably the middle part of a section of more than 23 m of 'Waterstones' recorded by Strahan (1882, p.11), the upper part of which is now nearly overgrown. Pollard (1981, p.558) recorded *Thalassinoides* at this locality. The sandstone in the above section caps a bold feature that can be mapped along the hillside immediately north of the marl pit.

A formerly large marl pit [525 741] east of Birch Hill Farm, between two north-south faults exposed, at the time of the survey, only a metre or two of well bedded, red, marly sandstone at the top of the south face of the pit. Due south of this pit, and between the same two faults, a cross-bedded sandstone more than 1.5 m thick, overlying 3 m of red marl, on red and green marl with thin sandstones, caps a high ridge between the two hospitals that were formerly Liverpool and Crossley Sanatoria.

Hondslough marl pit [538 731] and old pits by Waste Farm [537 734] are now nearly overgrown; in the westernmost of the latter pits, Strahan (in MS) recorded red and green shales (supposed 'Keuper Marl') on the east side of a fault. At Hondslough marl pit he recorded dips as high as 40° in marl and shale also referred to the 'Keuper Marl'. In view of the section proved in the Geological Survey Borehole at Hondslough (Appendix 1, p.102) these strata are here grouped with the Tarporley Siltstones.

The best exposure of many east of New Pale Farm [524 723] is in a small marl pit 400 m ENE of the farm where 1.2 m of red and grey sandstone overlie more than 2.5 m of soft red marl/mudstone. In a deep valley 550 m SE of the farm, sandstone of typical 'Waterstones' facies makes a 3 m cliff [528 720] on the east bank of the brook.

There are many exposures, probably low in the Tarporley Siltstones, in Ashton Brook where it flows through Ark Wood [522 714] and in old marl pits immediately to the north. These are mainly of gently inclined, red, sandy marl-mudstone with red and grey silty sandstone bands. More thickly bedded sandstone dips steeply east in a large marl pit [522 717] north of the wood and these exposures are believed to lie close to a major north–south fault.

The Tarporley Siltstones are well displayed in three faulted belts of country north and east of Kelsall and in the outlier of Pale Heights-Eddisbury Hill (Figure 10). In the westernmost faulted belt they make bold ridge-and-slack features between Lower Longley Farm and Kelsall village. The typical sandstone lithology is visible in a brook [520 682] SW of The Elms, and in road cuttings at Bank House [521 681] and SE of Rookery Farm [520 680]. Marls were extensively dug in the fields west and north of Grub Lane [524 685]. In the middle faulted belt, light grey shales with thin sandstones near the base of the formation were formerly dug in Kelsall between the church and Broom's Lane, and the same beds underlie the long slack above the steep scarp of Longley Wood. More sandy beds overlying these basal shaly strata make the bold ridge between Brock's Wood and Longley Covert. Old quarries [528 698] at Brock's Wood show:

| | Thickness m |
|---|---|
| Well bedded, hard, medium-grained, grey sandstone | 3.05 |
| Micaceous, grey, sandy shale | 1.37 |
| Hard, medium-grained, grey sandstone | 0.61 |

Pollard (1981, p.558) discovered *Planolites* and *Thalassinoides* hereabouts.

Higher beds are seen in the quarries [528 693] by Longley Farm:

| | Thickness m |
|---|---|
| Well bedded, red and grey sandstone with shaly partings and lines of small cavities | 3.5 |
| Red sandstone with many small cavities | 0.3 |
| Red shale and sandstone | 0.8 |
| Grey sandstone, the top part with many small cavities | 1.2 |
| Grey, micaceous shale with reddish streaks | 1.1 |
| Hard, well bedded and well jointed, medium-grained grey sandstone | 4.6 |

The sandstone at the base of this section is the same as that at the top of the preceding section and, according to Strahan (1882, p. 11), it was worked for hearth slabs and window sills. Higher marls with thin sandstones were dug [531 699] in the valley east of Brock's Wood, and the highest beds preserved in the middle faulted belt, mainly thinly bedded sandstones, have been extensively dug [533 693] on The Yeld (formerly The Heald).

In the easternmost of the three faulted tracts of Tarporley Siltstones in the Kelsall area, sandstones overlying the basal shaly beds make a bold ridge from Grey's Gate [536 704] to south of Lea Croft [534 698], and steep banks south of Yeld Croft [537 692]. There are several old marl pits in higher beds in Nettleford Wood, the best section [538 692], 180 m ESE of Yeld Croft, showing more than 6 m of well bedded, red, micaceous sandstone with marl-mudstone layers and bands of marly sandstone. In the outlying area north of Kelsborrow Castle, 3 m of well bedded red sandstone with layers of soft, red marl-mudstone are visible in an old marl pit [533 680] on the north side of the road at The Waste. A similar thickness of more marly beds is exposed [534 682] about 180 m to the NE. In the Pale Heights- Eddisbury Hill outlier (Figure 10), the base of the formation is visible on the north side of a steep-sided hollow about 400 m SW of The Old Pale farm. Here, 0.5 m or so of red marl-mudstone rests on typical 'Frodsham Beds' facies well displayed in the banks below. Higher beds, red and grey flaggy sandstones with partings of marl-shale, were formerly dug [543 696] near the water tower on Pale Heights and in a large pit [549 697] 400 m NW of The Old Pale. On Eddisbury Hill a marl pit [553 694] within the earthwork enclosure shows 1 m of well bedded silty sandstone on red and grey micaceous marl-mudstone, and similar beds were dug immediately north of the earthwork. A good exposure near the east

corner of The Old Pale farm buildings shows 2 m of thick, but well bedded, medium-grained, red sandstone with marl-shale partings resting on red and grey marl-shale and marl-mudstone with bands of silty sandstone.

*The centre (Willington to Tarporley)*

The near-horizontal Tarporley Siltstones outlier of Tirley Lane was extensively dug on both sides of the lane for 'marl' and stone. Up to 3 m of thinly bedded sandstones near the base of the formation are exposed [541 667] 200–300 m WNW of Higherbarn Farm. The northern part of the much larger outlier of High Billinge, Hollins and Quarrybank was likewise extensively dug for 'marl' and stone. Sections of 3 to 5 m of typical 'Waterstones' lithology are visible in a pit [553 666] 200–300 m NW of Hollins Hill Farm, in an old quarry immediately NW of High Billinge House, and in smaller exposures in many other pits between High Billinge and Holbitch Slack. The sections exposed in Holbitch Slack are noted on p. 28. In the western part of this outlier the large disused quarry [549 657], from which Quarrybank derived its name, exposes about 3 m of flaggy sandstone overlain by up to 9 m of shale and flaggy sandstone. It was probably worked mainly as a source of building slabs.

There are many fine sections in the large drift-free outcrop of Luddington Hill, Summerhouse Bank, Tarporley and Eaton. On Luddington Hill, 5 to 6 m of sandstone and shale are exposed in a quarry [567 648] NE of the summit of the hill, and in another [565 646] immediately west of the summit. There is also a succession of shorter sections in the valley head [561 645] west of the A49. A conspicuous section in Utkinton Road [550 642], north of Holly Bank, shows up to 9 m of typical sandstones with 'marly' layers, and a nearby scar [554 640], high on the southern slopes of Summerhouse Bank, exposes some 6 m of similar strata for a distance of about 200 m. A large quarry [546 634] alongside the A51 at Road Street, near Tarporley, exposes some 7 m of flaggy sandstone overlain by more than 5 m of shale and sandstone. According to Strahan (1882, p. 12), who listed 16 m of strata, the flags in this quarry were formerly 'worked up into window sills'. Strata at about the same horizon as those exposed in the Road Street quarry were also excavated in a quarry [547 641] at Ridgehill Farm.

One of the most conspicuous sections, showing many metres of characteristic 'Waterstones' lithology, is in a deep cutting [548 633] occupied by the A49 (Forest Road) at Portal, in the northern outskirts of Tarporley. The strata are near-horizontal and are exposed for a distance of nearly 200 m. Farther east, Eaton village is built on drift-free Tarporley Siltstones, and they are visible in many small laneside exposures. The very interesting fossiliferous section proved at Eaton Pumping Station is detailed in Appendix 1, p.101. Between Eaton and Eatonhill Farm [558 640] the hillside features illustrate typical 'ridge-and-slack' topography resulting from the alternation of hard sandy, and softer marly beds so characteristic of the formation.

*The south (Tiverton, Beeston)*

Tarporley Siltstones were encountered in the borehole at Gardenhurst [555 607] Tiverton, between the depths of 24.38 and 48.77 m and these gave a good supply of hard water (Wray and Earp, 1944, p.33).

The following section is seen 100 m NE of Castlegate Farm [541 593] Beeston, immediately east of the Peckforton Fault:

| | Thickness m |
|---|---|
| **DRIFT** | |
| Gravel with a few large sandstone boulders | 2.10 |
| Sand, loose brown with some quartzite pebbles | up to 1.83 |

SOLID

|  | Thickness m |
|---|---|
| Shale, red-brown with thin hard sandstone bands and bedding plane slickensiding | 1.83 |
| Sandstone, red-brown, fine-grained hard compact micaceous, with angular and rounded grains; rather broken and slickensided | 3.05 |

Both sandstone and shale are assigned to the Tarporley Siltstones; the sandstone was quarried formerly for roadstone. The beds dip at 10° due east.

*The north-east*

In the north-eastern part of the district, to the east of the outcrop, several boreholes prove the Tarporley Siltstones beneath the overlying mudstones. These include Crabtree Green water bore (Appendix 1, p.98) [5797 7084], made in 1934 for the Northwich Rural District Council, Austin's Springs water bore (Appendix 1, p.96) [5952 6722] for Winsford Urban District Council in 1899, and the Marston Borehole [6691 7540] made for the Salt Union in search of coal and other minerals in 1892 (Figure 12). Only drillers logs are available, and the most useful in terms of detail is the Marston Borehole. De Rance, who published the record (1895, pp.269–302) put the base of the 'Waterstones' here at 616.33 m, where beds described as 'red sandstone with marl' and 'red marly sandstone' rest with a sharp junction on soft sandstones of 'Frodsham' facies in the Helsby Sandstone. The driller's description of the beds above the junction is in accord with the lithology of the 'Waterstones' facies. There is, however, the same difficulty in precisely fixing the top of the Tarporley Siltstones as was experienced in the Stockport (98) district (Taylor and others, 1963, p.69) and for the same

reason — a gradual passage upwards into the overlying mudstones and a recurrence of the 'Waterstones' facies higher up the succession between 394.34 m and 447.85 m (see p.35).

The identification of beds below 440 m at Austin's Springs and 274 m at the Crabtree Green Borehole as Tarporley Siltstones, and not as an upper 'Waterstones' phase, is based on the interval between the estimated horizon of the base of the Northwich Halite and the top of the arenaceous formation at the two sites, and the assumption that there is no marked change in the thickness of these beds between here and the Marston, Plumley No. 3 and Booth Mill bores.

## LOWER MUDSTONE DIVISION (formerly 'Lower Keuper Marl')

The lower mudstone division contains all the lithological elements that are present in the Tarporley Siltstones, the difference being that the proportion of mudstone is greater in the higher formation. The outcrop of the division occupies some 70 km² of ground between the Overton and Winnington faults in the north-eastern part of the district and there is a narrow tapering outcrop west of the East Delamere Fault around Beeston and Tiverton.

The beds are poorly exposed for drift is everywhere thick. We rely on three boreholes, those at Austin's Springs, Crabtree Green and Marston, for our knowledge of the general stratigraphy (see Figure 12). The complete thickness of 445 m was proved at the Marston Borehole, sited a short distance outside the north-eastern corner of the district. This section shows a general decrease in arenaceous content from

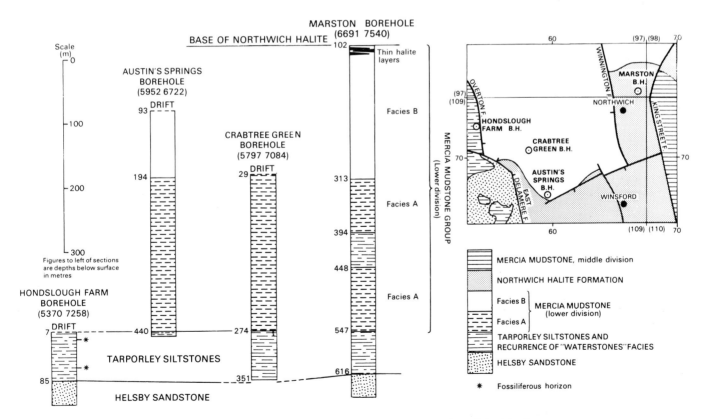

**Figure 12**  Borehole sections in the Tarporley Siltstones and the lower mudstone division of the Mercia Mudstone Group

the bottom to the top of the formation. The lower half may be described generally as laminated fine-grained dolomitic siltstone and mudstone alternating with structureless red (or red and green) silty mudstone; in the upper half the unstructured silty mudstone component is dominant and the laminated component subordinate. The two types of strata are shown as Lithology A and B respectively in Figure 12.

The recurrence of 'Waterstones' facies about 100 m above the base of the mudstone division in the Marston Borehole was noted by De Rance (1895, pp.269–291). He named the higher arenaceous phase 'Upper Keuper Sandstone' in this bore, and found the same beds in the Booth Mill Borehole [7682 7936] which he examined in 1902, though the record was published much later by Sherlock (1938, p.52). Sherlock (1921, p.47) also identified the same beds in Plumley No. 3 Borehole [7095 7534]; the term 'Upper Keuper Sandstone' must obviously be discontinued for these beds probably lie in the Scythian or Anisan stages of the Early and Middle Trias (Warrington, 1970b, p.200). It appears that the mudstone between the Tarporley Siltstones of the Chester district outcrop and the higher arenaceous phase of the Marston Borehole becomes more arenaceous north-eastwards, probably towards the edge of the depositional basin, eventually becoming, at the Ashley Borehole in the Stockport (98) district, indistinguishable from the lithology of the Tarporley Siltstones below the 'Upper Keuper Sandstone' (Taylor and others, 1963, p.69), so that there is an unbroken sequence of that facies at least 273 m thick, which may all be best referred to the Tarporley Siltstones.

In the north-eastern part of the district the Marston Borehole, and some of the exploratory boreholes of Imperial Chemical Industries Ltd., prove beds of impure halite up to 5 m thick in the 30 m of beds immediately below the Northwich Halite. Both the Southley Common Borehole [5816 5934] in the south of the district, and the Wilkesley Borehole [6286 4144] in the Nantwich (122) district proved mudstone with salt inclusions in this position, though there were no beds of salt. In view of their impersistence these beds are not included in the Northwich Halite, the base of which, following the practice in the Stockport and Macclesfield memoirs, is put at the base of the 'Bottom' or 'Hundred-Foot' Salt.

## Details

### Kingsley to Northwich

The main outcrop of the lower mudstone division lies between the Overton and Winnington faults in the Kingsley-Sandiway-Northwich area. Most of this ground is occupied by thick glacial drift deposits, and exposures are restricted to three small patches along the northern margin of the district and strips low down on the banks of the Weaver and its tributaries north of Winsford.

West of the Overton Fault near Kingsley a small area of lower mudstone [530 745] extends from the Runcorn (97) district to the north. There are no good exposures, and the presence of the mudstone is inferred from the soil content and from its relationship to the Tarporley Siltstones on the rising ground to the south. A small window through the drift [573 746] north-west of Crowton exposes a patch of lower mudstone which is part of the main outcrop east of the Overton Fault, as is an exposure in a gully [606 747] that runs northwards beyond the district west of Weaverham.

In the northern environs of Weaverham the valley side sloping down towards the high terrace of the Weaver is composed of the

lower mudstone division, patchily covered by boulder clay. The boundary between the two is inferred mainly from soil type but, at the eastern end of the exposure [626 744], a digging in the bank close to the river alluvium exposed red and grey mudstone. Mudstone is also mapped along the Barnton Cut, a navigation channel which follows the loop of the Weaver. East of this, and partly in the Runcorn (97) district to the north, mudstone is mapped along the north bank of the river Weaver as far as Anderton, where the Winnington Fault brings in the Northwich Halite to the east. South of the town there are no solid exposures between where the Winnington Fault crosses the Weaver [651 721] and the railway bridge [642 706], but between the railway bridge and the Hartford road bridge, some 600 m to the north-east, the boulder clay low down on the valley side is heavily laden with red mudstone debris, and it is likely that bedrock cannot be far below the surface. From the railway bridge southwards, mudstone is at surface along narrow strips of ground low down on one side or other of the river, either beneath boulder clay or below a terrace edge. The exposures extend for a short distance away from the main valley up several tributary gullies, the largest of which is The Riddings, a branching system that meets the main valley close to the west side of the railway bridge, and there is a smaller gully some 250 m to the south. A mere 2 m of mudstone are visible in the gully-bottoms beneath the boulder clay, the dip being gentle and variable around north-east. The lithology in the c17 m of strata encompassed by the exposures is red and grey banded mudstone with thin harder sandy bands.

Mudstone exposed along the east side of Valeroyal Cut between Valeroyal Locks [640 703] and where the Moulton Fault crosses the Weaver [648 691] is apparently structureless. Only two exposures in tributary gullies [6457 6993; 6475 6955] show the unweathered lithology which is in both cases red blocky mudstone. No estimate of the thickness of this structureless mudstone can be made because of the absence of measurable dips.

Two boreholes, at Austin's Springs and Crabtree Green, penetrated substantial thicknesses of the lower mudstone division in the outcrop area. They are correlated in Figure 12 with the Marston Borehole section which, though outside the district, is described here because it has been thought desirable to include all the Northwich part of the saltfield in this Memoir (see p.46). De Rance (1895, pp. 269–291) published the driller's journal of the Marston Borehole [6691 7540], and noted a recurrence of lithology resembling the Tarporley Siltstones some 99 m above the top of that formation (see p. 37).

At the Austin's Springs Borehole [5952 6722], sunk in 1899 for Winsford Urban District Council (Appendix 1, p. 96), Sherlock (in MS) tentatively placed the base of the 'Keuper Marl' at 440.44 m, and the base of the drift at 92.66 m, giving a proved thickness of 347.78 m for the lower mudstone. The driller described the rock as red marl to 140.4 m, red and grey sandy marl to 237.4 m, then mainly strong sandy marl to the top of the 'Waterstones'. Gypsum is recorded at intervals throughout. From the base of the drift to about 99 m the driller also recorded broken red marl with gypsum and, since the borehole is only 300 m from the mapped line for the limit of the salt subcrop, this could well represent the zone of collapsed strata fringing the salt crop, and would give some support for the position of the salt boundary as mapped, in an area where evidence is otherwise lacking.

The Crabtree Green water bore [5797 7084] was made for Northwich Rural District Council in 1934. The driller's log (Appendix 1, p. 98) was interpreted by D. A. Wray, who placed the base of the 'Keuper Marl' (top of the Tarporley Siltstones) at 274.45 m (see p. 37) at a level where the mainly argillaceous sequence gave place downwards to predominantly sandy 'Waterstones' lithology; the base of the drift being at 28.98 m, the lower mudstone sequence proved was 244.73 m thick. The borehole is 1.5 km away from the nearest part of the mapped limit of the salt, and may well be outside the zone of collapsed strata. At the time of survey (1953) some of the larger (41 cm) cores had survived near the site of the hole, and included red and grey sandy mudstone with ripple marks, salt pseudomorphs and gypsum in thin bands and veins, together with much red mudstone, completely weathered to clay. These cores are identified as part of the lower mudstone division because, according to the driller's journal, the hole diameter was reduced to less than 41 cm at 221 m depth, well above the top of the 'Waterstones' facies.

The lithological descriptions used by the drillers are different in each of the three boreholes that prove the lower mudstone division, and are depicted in Figure 12. In the Crabtree Green record an alternation of 'hard shaley marl' with 'soft red (or green) marl' clearly refers to beds similar to those, at about the same horizon, described in the Radbroke Hall Borehole [7651 7501] in the Stockport (98) district. These were used as an example of the rhythmic nature of Mercia Mudstone ('Keuper Marl') sedimentation (Taylor and others, 1963, p. 77). Likewise, alternations of 'red and blue sandy marl with gypsum' and 'red (or blue) marl with gypsum' in the Marston Borehole record may be taken to refer to similar beds. These are shown as Lithology A in Figure 12, and are more properly described as laminated fine-grained dolomitic siltstone and mudstone, alternating with structureless red (or red and green) silty mudstone. The upper part of the lower mudstone division, shown as Lithology B in Figure 12, is also composed of rhythmic alternations, but here the unstructured red (or red and green) silty mudstone component is dominant and the laminated siltstone/mudstone component subordinate. Overall, the upper half of the unit is more argillaceous than the lower.

In the north-eastern part of the district the Marston Borehole and some of the exploratory boreholes of Imperial Chemical Industries Ltd., prove beds of impure halite up to 5 m thick in the 30 or so metres of beds immediately below the Northwich Halite (see Figures 11 and 12).

*Beeston and Tiverton*

There are no exposures here, the whole area being thickly drift-covered. The presence of the lower mudstone division at outcrop is inferred from the fact that Tarporley Siltstones in the Tarporley area dip towards this ground at up to 18° with no evidence of an intervening fault, and a disused water bore at Gardenhurst [5556 6067] gives the following record:

|  | Thickness m |
|---|---|
| Drift (probably sandy boulder clay on loose sand) and 'Keuper Marl' | to 24 |
| 'Keuper Waterstones' to | 49 |

The outcrop is bounded to the east by the East Delamere Fault, the line of which is highly conjectural hereabouts.

## NORTHWICH HALITE

No one borehole or shaft has proved the whole of the Northwich Halite within the district. Figure 13 is a composite section based mainly on the Over Borehole, which proved all but about 17 m at the top and 15 m at the bottom of the sequence. The highest beds are taken from the Worleston Borehole and the lowest from the Meadowbank No. 3 shaft at Winsford. Figure 14 shows the correlation of key sections. Only three beds in the sequence have names, the Bottom Bed (Bottom Rock, Hundred-Foot or Main Salt), the Thirty-Foot Marl (Middle Marl) and the Top Bed (Top Rock). The names stem from the days when the two named salt beds formed the basis of a thriving rock-salt

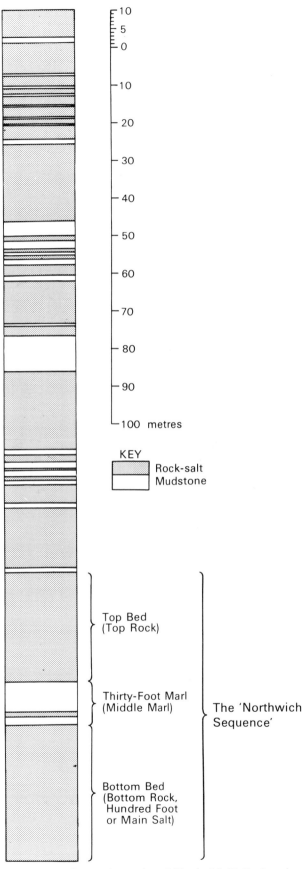

**Figure 13** Composite section of Northwich Halite based on the Over and Worleston boreholes and Meadow Bank No. 3 Shaft

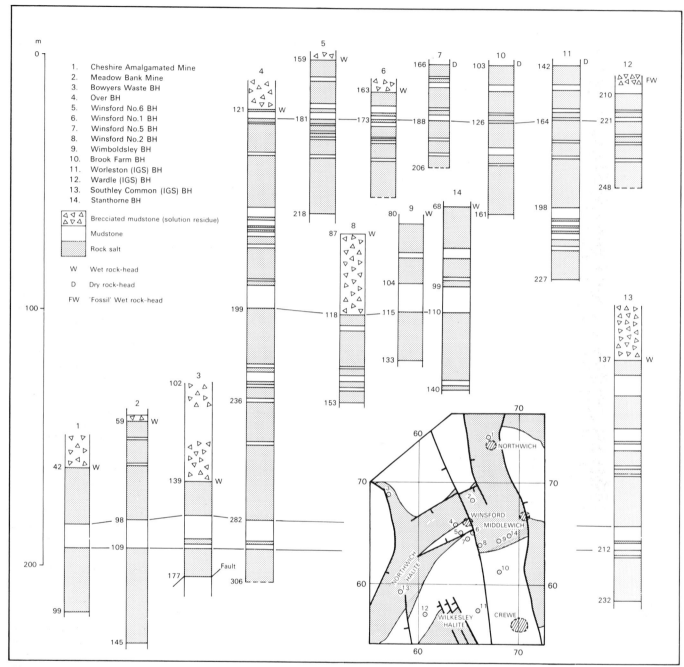

**Figure 14** Sections in the Northwich Halite

mining industry, first at Northwich and then at Winsford (see p. **84**). Imperial Chemical Industries Ltd., have divided the sequence into zones for ease of reference; their zones B, C and D correspond to the three named beds.

The total thickness of the formation is about 218 m west of the Winsford Fault and immediately south of Over. At Wilkesley, 25 km south of Over, it is 190 m. At Byley, 9 km ENE of Over, the thickness is 283.2 m; this suggests that east of the King Street Fault (Figure 20) the floor of the Cheshire Graben subsided more rapidly during sedimentation than to the west of that fault.

Correlations with the aid of microfossils have established the approximate time equivalence of the Northwich Halite and the Preesall Halite in west Lancashire, both being in the Anisian Stage (see Figure 11), although there are older halite formations than this both in the west Lancashire and the Irish Sea (Warrington and others, 1980).

Halite is the dominant mineral but clay and silt, distributed throughout the sequence both as inclusions in the salt and in beds of mudstone and silty mudstone up to 10 m thick, account for about 24 per cent by volume of the whole deposit. Other impurities include gypsum in small irregular masses and traces of the carbonates of calcium and magnesium, usually in the combined form dolomite.

The Bottom Bed was the main source of rock-salt for many years at Northwich, and is worked today at Meadowbank Mine, Winsford. It contains exceptionally clean layers, the run of mine rock at Winsford yielding upwards of 95 per cent sodium chloride. The bed varies in thickness from 35 m at Meadowbank Mine to little more than 20 m at the Southley Common Borehole. Some of the cleaner parts of the salt show light and dark bands varying from 25 to 100 mm in thickness. The bands have been regarded as annual layers (see for example Evans, 1970, p.115). Ward (1898, pp. 541–542 and plate) describing the banding in an unspecified Northwich mine wrote of rock salt striped 'almost like a zebra'. At Meadowbank Mine the bands exhibit a regular polygonal pattern which has recently been explained by Tucker and Tucker (1981, p. 495) as the result of thermal contraction when the halite was subaerially exposed after deposition. Other beds in the sequence also exhibit banding. Evans (1970, p.110) regarded banding as the primary state and structureless salt with irregular mudstone inclusions as the result of recrystallization of banded salt. In the structureless salt/marl beds there is every gradation from salt with around 10 per cent of silty mudstone, to a mudstone matrix with large isolated halite crystals.

The Thirty-Foot Marl includes a persistent bed of salt in its lower half in bores from Winsford westwards to Bowyer's Waste and southwards to Southley Common. Miscorrelations of this marl bed have been responsible for past misconceptions about the stratigraphy of the saliferous sequence. For instance, the mudstone between 104 and 118 m in the Wimboldsley Borehole (Figure 14, No. 9) was wrongly identified as the Middle Marl by Sherlock (1921, p.34) who thereby mis-identified the salt beds above and below it as Top Bed and Bottom Bed.

The Top Bed is the lower and cleaner half of a thick salt sequence of which the upper part is removed by solution in the Northwich area.

## NATURAL BRINE AND SUBSIDENCE

Solution of salt, seepages of brine and resultant subsidences in Cheshire are not recent products of Man's activity in the saltfield but are of geological antiquity. Ever since erosion removed the Upper Trias, Jurassic and possibly even Cretaceous cover to bring Triassic salt within the influence of mobile groundwater the process must have continued much as it does today, and the slow removal of salt by solution has gone on parallel with, and at about the same rate as, the general lowering of the land surface by erosion. Away from present and past centres of artifical brine abstraction the movement of brine and its replacement by fresh water is slow; the saturated brine layer protects the salt from further solution and there may be few records of subsidence. Moreover, the routes over the 'wet' rockhead by which the brine moves appear to become stabilised for long periods, and so villages and churches with long histories of stability exist in the 'wet' rockhead areas.

Brine seeping from the saltfield ultimately reaches the sea, and although its precise route is uncertain it seems likely that much of it flows along the deep gravel-filled channels that underlie the river valleys. In ancient times, before pumping lowered the level, some of the brine found its way to the surface in the low parts of river valleys, causing brine springs. These were some of the earliest sources of salt, and there are records of some of these 'wyches' in the Domesday Book. Sherlock (1921, p.51) cited records of ancient brine springs

outside the saltfield and close to the river Weaver at Hartford Bridge, Weaverham and Kingsley, and boreholes which proved brine at Barnton, Dutton and Frodsham. All these brines were weaker than the rockhead brine of the main saltfield, but some were strong enough to be used for salt-making. Their existence suggests a seepage along the Weaver valley to the Mersey estuary west of Frodsham.

The solution rate in the saltfield was probably accelerated temporarily when sea level was unusually low as in late-glacial times, and when surface streams, and probably the brine seepages also, were rejuvenated. Equally, high sea level would render the system sluggish, and movement might virtually cease over wide areas of rockhead. In this district at least, modern-day conditions where there has been no pumping appear to be closer to the latter condition than the former.

The earliest exploitation of brine was from springs occurring where natural seepages reached the surface either within or outside the saltfield. This presumably had no effect upon the natural regime of slow solution and subsidence for the brine would have run to waste in the rivers anyway. The discovery of copious saturated brine at the rockhead at Marbury (Sherlock, 1921, p. 3) transformed the industry. It also led to extensive damage from subsidence, for the ensuing large-scale pumping of the rockhead brine (known in the industry as 'wild brine') greatly accelerated the solution of the salt surface, particularly at the points of entry of fresh water, and the established pattern of natural brine movement in the vicinity was radically modified. It would be difficult to improve on the account of Thomas Ward (1900, pp. 246–247) of what happened.

'The level of brine was rapidly lowered, each pump became a centre towards which the brine travelled, and the fresh water travelled with equal rapidity to take its place. Instead of there being a gentle even flow over the whole surface of the salt-bed and a slight imperceptible lowering of its level, the fresh water cut channels on the salt in the direction of the pumping centres'.

'These channels widened and deepened until the water had taken up as much salt as it would contain, and then they died away as they approached the pumping centres. Where the fresh water first came into contact with the salt, the dissolution was the most rapid and extensive. As the level of the salt-bed was lowered the overlying marls followed, and the contour of the salt-bed was repeated on the surface of the overlying land. The marls, breaking and cracking in following the sinking salt, allowed more water to pass down and so intensified the subsidence'.

Subsidence consequent on the pumping of brine was not always close to the points of abstraction. The fresh replacement water soon became fully saturated with salt and was thereafter incapable of producing further subsidence; thus there was a tendency for the most distant parts of the induced brine streams to exhibit the most active subsidence, and for these active zones to migrate away from the pumping centres, in some cases forming branching systems.

During the resurvey in 1954–58 it became apparent that the topographic expression of subsidence due to brine abstraction was influenced by the type of drift deposit that overlay the solution zone. Where there is a continuous layer of boulder clay in which sand beds are thin or absent rather broad shallow depressions are formed, suggesting that fresh water percolates down towards the salt surface on a broad front, and not at particular points. Some depressions developed into troughs in the manner described by Thomas Ward (above) and by the outward migration of the fresh water ingress points. Similar linear subsidences (though in that case related to the stratigraphically higher Wilkesley Halite) were described by Evans (Evans and others, 1968, pp.145–148) in the Elworth and Sandbach areas to the east of this district, together with a detailed explanation of the mechanics of their formation. The linear subsidences mark the course of brineways on the surface of the salt, reflecting in some instances the strike of individual salt beds at the subcrop. Lakes or 'flashes' formed along many of the linear subsidences, but it is doubtful if much of the impounded water can have assisted in the solution process because of the underlying seal of boulder clay.

Thick sand beds, such as the Delamere Sands, provide large reservoirs of fresh water, and here the initiation of brine pumping had a dramatic effect. The lower surface of the sand is in places deeply channelled into collapsed strata overlying the 'wet' rockhead, cutting through any intervening boulder clay layer. The percolation routes through the mudstone to the salt are shortened beneath the channels and are therefore preferred, the effect being to concentrate the access of fresh water to a few restricted points. Here solution cavities rapidly developed, the roof being further loosened, letting in more water, so that eventually a wholesale rapid collapse of mudstone and overlying sand into greatly enlarged cavities caused conical 'crater' subsidences at the surface. Descriptions of these are given on pp.50–51).

Linear subsidences (Calvert, 1915, p.304) in this district are best exemplified south of Winsford between the south end of Bottom Flash and the confluence of Ash Brook with the river Weaver [6692 6313]. Details are given on p.52 (see Figure 16). Some of these features show a strong ENE–WSW linearity, and the evidence from several boreholes in the vicinity is that this is also the direction of strike. The features may, therefore, correspond to the subcrops of individual salt members at the 'wet' rockhead. The marl partings at rockhead are likely to stand proud between the dissolving edges of the salt layers, as shown diagrammatically in Figure 15, and to form interfluves between brineways. Linear subsidences frequently develop outwards from the pumping centre, the most active part is the distant end where fresh water gains access through new fractures; the nearer parts become less active because the brine is already saturated when it reaches them.

South of Bottom Flash the brineways beneath the linear subsidences are feeders to a major brine seepage flowing northwards down the Weaver valley. This was once the source of raw material for a score or more of brine pits in the town. The smaller flashes are caused by the warping of the valley bottom where it is crossed by one or more linear subsidences. The large Bottom Flash is where a number of brine streams converge as they approach the old pumping centres at Winsford, and are supplemented by fresh water introduced through continuously rejuvenated subsidence fractures along the valley sides and beneath the lake. The result has been the severe collapse of a wide strip of the valley floor. The linearity of this feature is related to the Weaver valley; it is not a linear subsidence in the same sense as are the tributary brine runs higher up the river, or those described by Evans and others (1968, pp.145–148).

Crater subsidences (Evans and others, 1968, p.145) in this district are confined to the region of thick Quaternary sands north-west of Winsford, and they extend in a well-defined belt on the south side of the Moulton Fault from Marton Hole east-north-eastwards towards the site of the old Newbridge Salt Works. Details are given on p.50. Certain deductions can be drawn from verbal descriptions of the way some of these subsidences began, obtained by Dickinson (1882, p.94), the events then being within living memory. The initial collapse at Marton in 1863 was sudden and unexpected. Dickinson likened it to the 'crowning-in' of flooded salt mines at Northwich (see below). Repeated enlargements of the crater occurred in the next few years, each heralded by the disappearance of the standing water in the hole. During this time the volume of sand that had subsided, accepting Dickinson's dimensions for the resultant funnel-shaped hole, must have exceeded 50 000 m$^3$ and this gives the measure of the solution cavity at rockhead into which the waterlogged sands flowed.

Evidence of direct hydraulic connections between points along the brine-run is given by the fact that an upward eruption of brine occurred on the north bank of Pettypool Brook simultaneously with a new and sudden collapse half a mile upstream. Dickinson recorded that at the same time there was also a check to the supply of brine to the Newbridge brine pit 'as though the run had become choked or diverted'. One envisages, therefore, a rather wide and open brine 'main' running along the line of crater subsidences, large enough to accommodate quite massive inflows of solid debris.

The line of the Marton-Newbridge crater subsidences is determined by the Moulton Fault bounding the saltfield to the north-west and the edge of the thick water-bearing sands to the south-east. Shearing stresses near the fault between the stable ground to the north and the collapsed ground to the south may have helped to open up the rock above the salt to the fresh water stored in the sands. The base of the sand is known to be highly irregular, and the sites of the crater subsidences may be where it is most deeply incised into the strata overlying the subcrop.

The southern branch of the Marton-Newbridge subsidences whose southern tip is west of Knight's Grange [640 671], occurs where rather thin water-bearing sands overlie collapsed beds, and are overlain by boulder clay. Instead of the massive funnelling-in of sand and water at relatively few points as on the main arm, the effect has been more diffuse, resulting in a broad depression flooded in places with a generally northerly flow of brine towards the main arm.

Now that the abstraction of natural ('wild') brine has virtually ceased in favour of the controlled method of pumping (see p.84) the rate of subsidence in the worst-affected areas has slowed down, but it has not stopped altogether. Dickinson (1882, p.96) noted that subsidence near a salt works at Nantwich was still continuing 25 years after brine abstraction had ceased, and in the Winsford area fresh subsidence scars are still (1980) forming in the fields bordering the Bottom Flash some 20 years after most of the town's brine pits closed down. It is doubtful if the movement of natural brine will ever quite return to the regime which prevailed before pumping began. Like any other modified drainage system, the artificially-engendered brine-runs have become entrenched in their new channels, and the rate of solution will be higher in the affected areas than elsewhere, at least for some decades.

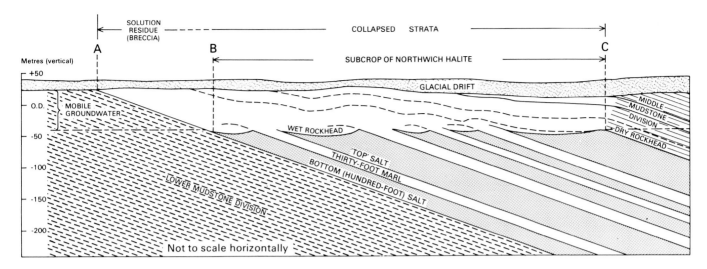

**Figure 15**  Diagram showing the relationship of 'Wet' and 'Dry' rockhead to the subcrop of the Northwich Halite

The area of potential subsidence resulting from solution at the salt surface is co-extensive with the mapped subcrop area (see pp.88, 91). The top of the salt in the subcrop area has rarely been seen *in situ*. Figure 15 summarises information about it, and the effect of its solution on overlying beds, from various sources. Numerous cored boreholes have proved broken and collapsed strata above the salt, and it is not uncommon for a cavity to exist at rockhead. The few observations of the upper surface of the salt suggest that it is very uneven and deeply furrowed by solution channels. Many old records refer to a layer of granular marl, taken to be an insoluble argillaceous residue from dissolved salt; by its nature it has rarely been recovered from cored boreholes. Dickinson (1882, p.79) whose information was from brine-shaft sinkers described the top of the salt thus:

'From what has been seen and is known of them, the spaces which a brine-run makes between the rockhead and the marlstone may be at first as thin as a sheet of paper, but the spaces become larger, and with the wearing away or solution of the rock-salt a peculiar structure of granular marl called 'horse-beans' ensues between the rockhead and the overlying bed of marlstone called the 'flag'. In this granular structure freer course is afforded for the flow. Spaces, at first only the size of rat-holes, become so large that a man can enter, and they increase into large caverns, and ultimately the ground about them subsides'.

The 'horse-beans' and brecciated strata, in conjunction with the rockhead voids, act as an aquifer for the brine. Saturated brine contains 26.8 per cent NaCl and has a specific gravity of 1.0255. The natural rockhead brines in the Winsford and Northwich areas have a salt content close to this figure, some of the shortfall being taken up by traces of other metallic salts. Brine therefore gravitates to the lowest parts of the aquifer, forming a layer beneath any fresh water which may be present. As the salt is dissolved and carried away as brine it leaves behind insoluble inclusions from the salt beds as well as the beds of stratified mudstone, some of them more than 10 m thick, which alternate with the salt layers, and a zone of residual material is created at the outcrop in the stratigraphical position of the salt/mudstone sequence.

The term solution residue, if strictly applied, would refer only to the insoluble material released from the salt beds—the 'horse-beans' of the old shaft-sinkers—and not to the beds of stratified mudstone. The practical difficulty is that as solution proceeds, with piecemeal sagging of the overlying measures, all stratification is lost. Bedded mudstone becomes brecciated and inextricably mixed with that released from the salt beds. The estimated thickness of this deposit in the Winsford area is about 53 m; this is the sum of all the mudstone beds in the Northwich Halite, together with the insolubles in the rock salt, estimated at 5 per cent of the salt thickness. It is convenient to refer to all this material as the solution residue of the Northwich Halite (see Figure 15).

The only exposures of the solution residue known in the district are of unstructured argillaceous material beneath boulder clay along the east side of the river Weaver between the conjectural position of the Moulton Fault [6486 6916] and the occurrences of well-stratified banded mudstone and siltstone [e.g. 6540 6851] opposite Meadowbank Mine. Boreholes RM 7 [6422 6882] and RM 10 [6428 6868], along the strike to the west, proved thick sequences of brecciated mudstone, interpreted as solution residue.

The mudstones overlying the subcrop are seen at outcrop in the Spurstow area, in the banks of the Gowy and the Shropshire Union Canal near Tilstone (p.50) and in the Weaver valley both north and south of Winsford. In all these areas it is presumed that these exposures are of the middle mudstone division, draped across the dissolved edges of the salt. The 'dip' is highly variable but the bedding is preserved. These strata are out of place, and they are called collapsed beds to distinguish them from the solution residue. Boreholes through the collapsed beds generally prove a sharp junction between them and heavily shattered and brecciated ground immediately overlying the salt.

'WET' AND 'DRY' ROCKHEAD

Figure 16 shows contours on the upper surface of the salt formation (rockhead) in the Winsford area. Parts are left blank for want of evidence, and other parts depend on extrapola-

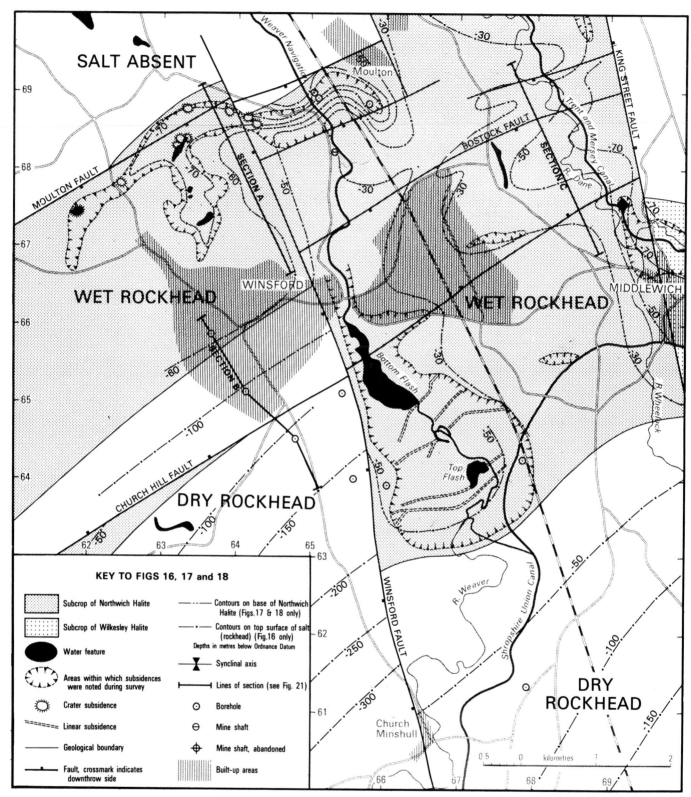

**Figure 16** Salt rockhead contours on the Northwich Halite around Winsford

tion; nevertheless, some useful conclusions can be drawn from it.

In the subcrop area (Figure 15) the top of the salt is referred to as 'wet' rockhead because this is where brine is to be found. Where the surface on both sides of the fault is 'wet' (see Figure 21c), the level of this surface is hardly affected by faults. Brine channels also traverse the 'wet' rockhead apparently little affected by the underlying structure. The Marton-Newbridge brine channel crosses the Winsford Fault north of Wood Farm [641 685], the slight deflection of the main channel at the fault being due to the displacement of the saltfield edge. Between Wimboldsley and Moulton, on the divide between the rivers Weaver and Wheelock, the undulations of the salt surface appear to bear little relationship to the tectonic structures which affect the salt beds. By contrast, the 'dry' rockhead area, where the salt top is below the level of groundwater circulation, behaves, so far as is known, like any other formation top. The boundary drawn between 'wet' and 'dry' rockhead is highly conjectural, and it is possible that the 'dry' area as mapped may carry solution channels, particularly near its up-dip edge, which are not now active but date from pre-glacial times when there was no protecting drift cover, or when there was a greater fall to sea level than at present.

The pattern of surface subsidence features in the Marton, Bradfordwood and Bark House area indicates the presence of a brine-gathering system which once fed to an abstraction point at Newbridge, and this is supported by historical records (see p. 50). Some idea of the depth of the rockhead channels is given by RM 29 Borehole [6359 6767], drilled slightly off centre along a main southern tributary of the system. Here rockhead level is 67.97 m below OD. It is therefore assumed that the deepest part of the channel is probably more than 76.2 m below OD, and on Figure 16 a contour for this value is drawn around the channel's assumed position.

There is no information on rockhead level in the subcrop area immediately west of Over. It may be assumed to undulate a little above or below – 61 m OD. At Southley Common Borehole [5816 5934], some 8.5 km SSW of Over along the subcrop, the salt rockhead level is 85.6 m below OD. At the Bowyer's Waste Borehole [5686 6881], 7.5 km WNW of Over and hard up against the East Delamere Fault which bounds the saltfield, rockhead is 60 m below OD, and at the Oakmere Borehole in the same vicinity [5768 6780] it is 64.6 m below OD.

South of Winsford and east of the Winsford Fault, the Weaver flashes and associated linear subsidences indicate the existence of another main brine-gathering system which fed the long abandoned abstraction points in the Winsford town area. There is little firm evidence regarding the present form of the rockhead in this part of the Weaver valley, and it has not been found possible to draw detailed rockhead contours here. It is not even clear where the deepest parts of the rockhead channels are. The Wimboldsley [6784 6425], Winsford No. 2 [6604 6392] and Ways Green [6554 6570] boreholes are all near the periphery of the unstable area and not near its axis, which probably lies closer to the line of the river. The level of – 75.6 m proved at Winsford No. 2 may indicate a falling-off of the salt top towards the Ways Green Fault, as happens also near the King Street Fault north of

Croxton Bridge. The Runcorn Alkali Borehole [6522 6671] proved the salt top at – 30.8 m at the beginning of this century, but the combined effect since then of abstraction at many brine pits along the Weaver valley at Winsford may well have altered the shape of the rockhead channels. Records of subsidence suggest that there was considerable solution of salt in the town area north of Winsford Bridge.

West of Middlewich the Fault brings the 'wet' rockhead of the Northwich Halite into contact with that of the stratigraphically much higher Wilkesley Halite. Evidence from the deep boreholes of Imperial Chemical Industries, Ltd., has refined the plotted position of the King Street Fault and now makes it likely that several old brine wells in Middlewich (Nos. 1 to 5 in Sherlock, 1921, fig. 5, p.36) are in the Wilkesley Halite, and not the Northwich Halite as was shown on the current geological map of the Macclesfield (110) district.

There is borehole evidence that the 'wet' rockhead has been dissolved in an irregular fashion along the west side of the King Street Fault between Stanthorne and Whatcroft to form what appear to be two separate 'deeps', one near Croxton Hall and another about 1 km farther north, separated by a low rockhead spur near Dairyhouse Farm. It is thought that most of the solution which caused these hollows occurred in pre-glacial times. The surface above the hollow near Croxton is affected by recent subsidences, which suggests that brine has flowed freely across the King Street Fault to nearby abstraction points in Middlewich. The amount of post-glacial subsidence, however, does not nearly match the depth and volume of the rockhead hollow. Moreover, the modern subsidence does not lie directly above the deepest part of the rockhead hollow, but a few hundred metres to the south-west. Brine abstraction at Middlewich probably reactivated solution in the Croxton hollow which had been protected since pre-glacial times by thick Quaternary deposits. There are no signs of modern subsidence above the more northerly rockhead hollow.

## MAPPING OF SALIFEROUS BEDS

The solubility of rock salt presents special mapping problems which, together with the conventions adopted to overcome them, are outlined in the Stockport and Macclesfield memoirs (Taylor and others, 1963, pp.78–79; Evans and others, 1968, pp.144–145). In Cheshire the salt beds can exist no nearer to the surface than the base of the zone of mobile groundwater. Instead of mapping an outcrop, therefore, we map a 'subcrop' where undissolved salt abuts against the base of the solution zone. The mapped subcrop is co-extensive with the area of 'wet' rockhead (see p.43) and both coincide with the zone of potentially active salt solution subsidences.

Figure 15 is an idealised section across the salt subcrop, the mapped boundary lines of which pass through B and C. Both of these lines are almost impossible to place accurately even with the aid of boreholes. The 'wet' rockhead surface is undulating, and in places deeply channelled by solution; therefore both the outer limit of salt (B) and the down-dip edge of 'wet' rockhead (C) are likely to be sinuous, especially where the rocks are gently inclined. The lines as mapped are everywhere conjectural, and are shown conventionally as smooth curves.

In mapping the saltfield the two prime sources of information are boreholes and the surface effects of solution-subsidence. The mapping convention adopted makes use of both these sources in the most practical way. Surface exposures of rock are of limited value in most of the saltfield because at best they are seen through a few narrow windows in the almost ubiquitous glacial drift deposits, and at worst nothing at all is seen. This is because it is in the nature of the salt subcrop that, being the site of long-continued subsidence, it tends to occupy low ground, where the glacial drift and alluvial deposits are at their thickest. The only line which might be mapped by surface exposures, even if these were adequate, would be that indicated by A in Figure 15, that is, the junction of the undisturbed lower mudstone division with the overlying salt solution residue. This line would not define the edge of the saltfield, which might be as much as several miles away, and would be of no practical value for either economic or planning purposes.

The attention of those who collect for stratigraphical purposes is drawn to the anomalous position of the outcrops of mudstone within areas mapped as Northwich Halite Formation. This fact might have been considered as of little importance when 'Keuper Marl' was thought to be unfossiliferous, but following recent advances in Triassic palynology this is no longer the case. Collectors must, therefore, take care that specimens from formations overlying, or even some distance up-dip of, saliferous beds are attributed correctly to their lithostratigraphical position.

## Details

### The Northwich area

R. L. Sherlock (1921) summarised all the information available in his time of the geology and salt industry of the Northwich area, and gave full references to previous works such as the seminal papers of G. W. Ormerod, Joseph Dickinson and Thomas Ward, various government papers dealing with saltfield legislation, and the compilative work of A. F. Calvert. These accounts were written before the correlation of the Northwich stratigraphical sequence with that of the rest of the Cheshire saltfield had been established, and no modern sheet memoir has so far dealt with the Northwich area as a whole because it lies in the corners of four geological 1:50 000 sheets. The opportunity is therefore taken now to describe the ground between the Winnington Fault and the Wincham-King Street faults from Budworth Mere and Pickmere southwards to Davenham and Whatcroft. The area includes the sites of all the Northwich mine shafts and brine pits, as well as most of the ground that was affected by subsidence through brine abstraction at Northwich.

Both Ward's (1898) and Sherlock's (1921) accounts had the benefit of the Marston Borehole record (De Rance 1895) which showed that the salt beds did not lie near the base of the 'Keuper Marl' as Hull (1869, p.82) had thought, but a considerable distance above it. The Marston record did nothing however, to counter the prevailing view that the Northwich sequence of Top and Bottom beds with the intervening Middle Marl constituted the whole Cheshire succession. Correlation with more complete saliferous sequences obtained from boreholes in other parts of the saltfield, the earliest of which were those of Imperial Chemical Industries Ltd., in the Holford Brinefield east of the King Street Fault, showed that the Top and Bottom beds of Northwich were merely the lowest two thick salt layers in a sequence which when fully present could be as much as 285 m thick, as for instance at Byley (Evans and others, 1968, pp. 277–279).

The earliest rock-salt mine shafts at Northwich found only the Top Bed. Only in 1781 was the Bottom Bed discovered as a result of exploration following the discovery of more than one salt bed at Lawton, near the eastern edge of the saltfield some 23 km to the south-east. Also, no records have survived from many of the older mines and brine wells. The result is that from upwards of 120 known sites of mines and boreholes in or near Northwich only 34 have provided firm information on the base of the Bottom Bed, that is the base of the main salt sequence, at Northwich. That horizon is chosen as the basis of structural contours in Figure 17 because all the salt workings lie above it, and it is, therefore, unaffected by subsidence.

In the immediate town area there are enough reliable records based on shaft sinkings and one cored borehole to allow the −75 m OD contour to be plotted on the base of the Bottom Bed. This reveals the presence of a very gentle syncline with its axis trending north-north-eastwards through the town centre between Leftwich and Marston. The lowest recorded points along the axis are 83 m below OD at Platt's Hill No. 2 Shaft and at Nelson Mine (Nos. 15 and 12 in Figure 17). Platt's Hill Mine suffered what had become a standard sequence of catastrophic events for mines in that part of Northwich (see Calvert, 1915, pp.213–215, 351–367). The Bottom Bed was worked for rock salt until 1880, then water from flooded mines to the north broke through a thin barrier of salt into the workings and the mine was abandoned. The shafts collapsed in 1893, and progressive subsidence over the mine coalesced with that at mines to the south and north, eventually to form the great lake known as Ashton's Flashes. References to accounts of the Northwich subsidence are given on p.84.

Mines were clustered along the synclinal axis, presumably because the greatest thickness of Top Rock was preserved there. The Top Rock thickness within the −75 m line is generally greater than 20 m, with maxima in the vicinity of the Marston Borehole (11) were it is 27.58 m. Outside the −75 m, line, Top Rock thicknesses are generally less than 20 m. At the Marbury Hall Borehole (2) it is 18.8 m, and at the Pickmere Borehole (5) 11.89 m.

The −75 m contour is shown as closing between the Marbury and Pickmere boreholes, but this is conjectural. There is no evidence of sharp folding, and the base of the Bottom Bed beneath Budworth Mere and Pickmere may be not much less than 75 m below sea level. Some 300 m north of Budworth Mere, a borehole [9527 7725] proved a mudstone sequence, with traces of salt between 88 and 100 m that probably represent one or other of the thin salt beds stratigraphically below the main Northwich Halite. There is no mention in the log of the brecciation which would certainly have been present if the bore had been located in the zone of collapsed mudstone and solution residue which fringes the normal salt outcrop (see Figure 15). The strong inference is that the Northwich saltfield is abruptly terminated to the north by a fault running nearly east–west, close by the northern edges of Budworth Mere and Pickmere, with a southerly downthrow of 55 m or greater. Such a fault would be comparable in its relationship to the major faults to west and east with the Moulton, Bostock and Church Hill faults in the Winsford area (see Chapter 5, Structure).

East of the synclinal axis, the log of Penny's Lane mine-shaft (24) records a Top Rock thickness of 17.22 m and at Wadebrook Mine (26), farther east still, it is reduced to 12.74 m. At Wincham Lane No. 1 Borehole (14) the correlation is not certain, but it seems that the Top Rock is either absent or much reduced.

Sherlock (1921, p.25) located the Wincham Fault at Wincham on the basis of boreholes, and showed its extension as passing between Bowman Thompson's No. 2 Brine Shaft (27) and the mine-shaft (28). Sherlock's line is here continued to pass close to the east of the major subsidence features south-east of Billingegreen, and to rejoin the King Street Fault at Whatcroft. The Broken Cross Borehole (32), where the Bottom Bed base is at 84 m below OD, is considered to lie between the Wincham Fault extension, with a modest downthrow east of around 25 m, and the King Street Fault with an easterly downthrow in that area greater than 500 m.

In the southern environs of the town, away from the main concentrations of rock-salt pits and brine shafts, two boreholes allow an estimate of the level of the Bottom Bed base; these are the Witton Junction (31) and Dane Meadows (30) boreholes. The Witton Junction Bore was made for the Widnes and St. Helens Salt Syndicate in 1890, and the salt proved between 75.59 m and 93.97 m is regarded as the Bottom Bed. Sherlock (1921, p. 20) placed the drift base here at 52.19 m depth, only 23.4 m above the salt top. This probably accounts for the diminished thickness of the Bottom Bed, 18.38 m, against a general thickness of around 27.5 m in the Northwich area. The Top Bed is missing. At the Dane Meadows Bore a salt identified as Bottom Bed was reached at 75.29 m, and the bore ended at 82.91 m still in salt. The base of the drift here, according to Sherlock (1921, p.19) is at 65.84 m, only 9.45 m above the salt top. It is impossible on the evidence available to estimate how much of the Bottom Bed has been removed by solution, and an arbitrary level for its base, taking a similar thickness of remaining salt as at Witton Junction (see above), is placed at about 80 m below OD.

Pimlott's Boiler Yard Borehole (29) is difficult to interpret. A record of 'red and grey marl with rock salt' between 74.22 m and 91.90 m is insufficient as a basis for correlation. The drift base was put by Sherlock (1921, p.19) at 62.03 m, only 12.19 m above the saliferous marls. One may suspect that in the case of this bore, as also at the Riversdale Bore and the two Leftwich bores (see below), drilling techniques when they were made were incapable of producing consistently reliable results in collapsed ground underlain by salt beds.

Sherlock (1921, p.27) stated that there were two boreholes at the Riversdale (35) site, made by different engineers, and differing in their details. Leftwich No. 1 (33) Borehole proved salt, but positive correlation of it is impossible. At Leftwich No. 2 (34), a salt bed between 94.18 and 98.76 m is thought to be the Bottom Bed, removed by solution except for 4.58 m, and with its base 71.63 m below sea level. A 2.23 m salt bed with its base at 106.98 m is correlated with the higher of two thin salt layers below the Bottom Bed at the Marston Borehole, (see Figure 12). A general conclusion is that the Riversdale and Leftwich boreholes lie in an area of abnormally thick drift, and the salt has been dissolved to a greater depth below surface than elsewhere. The salt beds met with in the bores are either remnants of the Bottom Bed or correlatives of the thin salt layers known to occur in the mudstones below the Northwich Halite.

West of the town centre the Baron's Quay shaft (22) is similar to the Marbury Borehole (2) both in the depth to the Bottom Bed base at 70 m below OD and the thickness, 18.42 m, of the Top Rock. At Winnington Hill No. 1 Brine Shaft (21), some 350 m away from the Winnington Fault, the Bottom Bed base is, at 62 m below OD, only 8 m shallower than at Baron's Quay, but the Top Bed is absent, and the inference is that drift thickness is much greater at Winnington Hill, although both bores are within 400 m of the river Weaver.

East of Davenham the – 100 and – 125 m contours on the base of the Bottom Bed show that there is a pronounced dip westwards and north-westwards away from the Whatcroft area. This is in sharp contrast to the structure around Northwich and Leftwich to the north. The two areas are thought to be separated by a fault, the line of which is drawn roughly parallel to other minor faults that trend at right angles to the major north-south fractures in the region, to pass north of the major Billingegreen subsidences and south of Davenham.

Calvert (1915, pp.215–216) referred to unsuccessful attempts to find salt along the Weaver valley west of the Winnington Fault, at Acton Bridge, Dutton Viaduct and at Anderton. On the principle that the records that survive tend to be the successful ones, it is like-

ly that many more abortive attempts by speculators took place, leaving no written records, but reinforcing the local unwritten lore that no salt exists west of the Winnington Fault—a view which agrees with the surface evidence. The latter includes the fact that there are no subsidences attributable to salt-solution west of the line in the Anderton area, as there are to the east of it; the two-mile long exposure of Mercia Mudstone (lower division) mapped by W. B. Wright along the north bank of the Weaver between Weaverham and Anderton ends abruptly at the fault line.

It has been noted elsewhere in the saltfield that solution subsidence may be severe close to faults, probably because water can percolate downwards to the salt more readily there, and the subsequent collapse is assisted by fracturing of ground close to the fault. As stated above, Budworth Mere and Pickmere are thought to lie along the northern edge of the saltfield where it is bounded by a fault. There are few references to the lakes in the geological literature. Sherlock (1921) refers to subsidences 'on the north side of Budworth Mere and near Pickmere' as evidence that the saltfield extends farther north than Rigby's Borehole (No. 4 on Figure 17), the most northerly of the Northwich group of provings, but he makes no suggestion that their existence owes anything to artificial brine abstraction. These meres pre-date that activity, as does Rostherne Mere which also lies at the periphery of the saltfield in the Stockport (98) district some 9 km to the north-east (Taylor and others, 1963, p.81). It seems probable that all three meres are the product of natural salt solution subsidence in the recent geological past, and certainly since the retreat of the last ice-sheet, as it is the drifts that show the surface effects. A possible explanation of the slowing down of subsidence at these meres, where it has probably been quite vigorous in the past, it that natural brine-runs, active during the immediately post-glacial low sea-level, are now inhibited by the present reduced fall to the sea (see p. 41).

The pronounced subsidences that run almost parallel to the railway line and canal near Billingegreen are thought to indicate where salt within the zone of solution is overlain by broken ground associated with the Wincham Fault. Much discussion has arisen as to whether the subsidences were due to pumping at Northwich or Middlewich, or at both places (Sherlock, 1921, p.63), some of it based upon a misunderstanding of all the geological relationships. It is possible to envisage a brine route 5 km long from Wharton southwards, close to the west side of the King Street Fault, crossing the fault near Middlewich where the 'wet' rockhead of the Northwich Halite to the west is in contact with that of the Wilkesley Halite to the east; the route northwards to the Northwich pumping centres is, however, little more than 2.5 km long, and geologically more direct. In the latter event, the most likely path for the brine from Billingegreen to the abstraction points at Northwich would be by way of the buried channel of the river Dane. The abrupt termination of subsidence at the fault-line near Billingegreen may reflect the fact that thick salt in the main Northwich Halite sequence with a continuous 'wet' rockhead is present south of the line, whereas to the north of the line there are only thin salt layers proved in the Leftwich bores.

Subsidences in the Northwich town centre are the result of the industrial exploitation of the salt, and have no structural significance, except that they tend to be located chiefly along the synclinal axis because the salt sequence was thickest there. Ashton's and Neumann's flashes, and the other smaller subsidence lakes in the town area are the present-day remnants of much wider effects in the built-up area of the town and along the river Weaver. These are discussed elsewhere.

### West of the Winsford Fault

The subcrop west of the Winsford Fault is mapped as a continuous belt 3 or 4 km wide from north-west of Winsford to Spurstow in the south-west, with a northern branch in the Oakmere area representing a hypothetical westerly downfold against the East Delamere

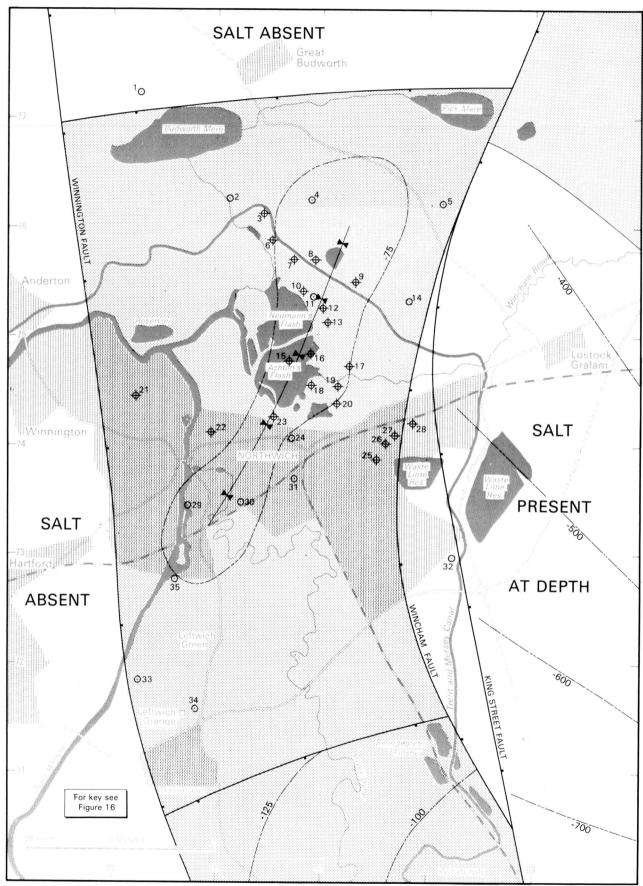

**Figure 17**  Geology of the Northwich salt district

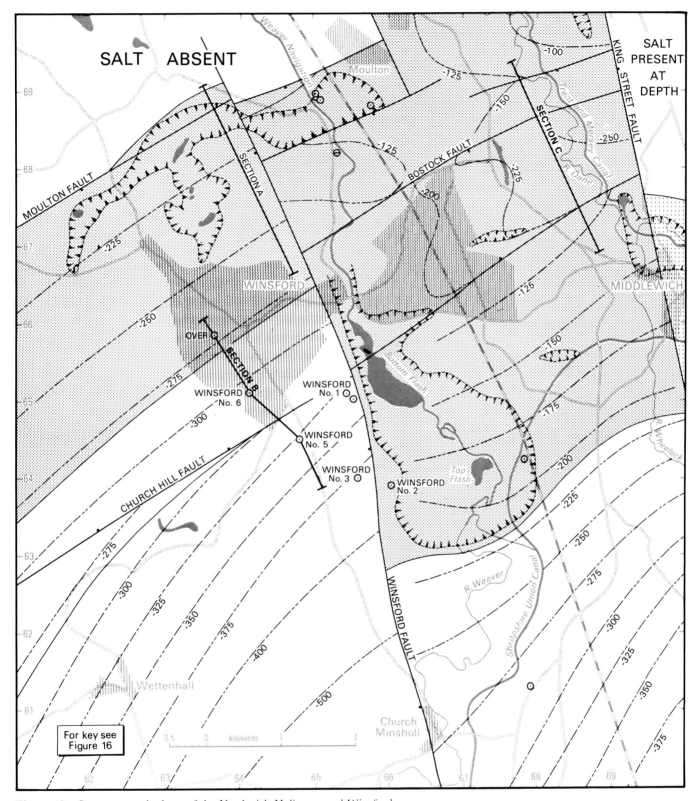

**Figure 18**   Contours on the base of the Northwich Halite around Winsford

Fault. There is inadequate geological control, either from surface evidence or from boreholes over most of this subcrop area, and the map is in places highly conjectural. Except for a few exposures along stream channels near Spurstow and Bunbury the solid rocks are completely obscured by thick glacial drift. There are undoubted surface subsidence features west and north-west of Winsford and around Oakmere; these are described below. In the neighbourhood of Little Budworth surface depressions some 200 and 500 m south of Old Hall [604 656] are less certainly attributed to solution subsidence, as are depressions west of Oaktree House [592 628]. Throughout the rest of the outcrop it must be assumed that hydraulic conditions over the salt surface are quiescent, and that there is no significant access of fresh water to the salt beds. Of 11 boreholes drilled to prove identifiable horizons in the saliferous beds, ten are in the northern part of the subcrop and eight are near Winsford. The greater apparent structural complexity in the Winsford area (see Figure 18) is due entirely to the concentration of information there, particularly from boreholes. From Little Budworth to the southern edge of the map only the Southley Common Borehole [5816 5934] proves the subcrop, and it is virtually certain that the map here is greatly oversimplified.

South of the Moulton Fault and immediately west of the Winsford Fault north-west of Winsford four exploratory boreholes made for Imperial Chemical Industries Ltd., proved 'wet' rockhead, solution having removed a thickness from the top of the saliferous beds varying from 16.15 m (RM 30) to 55.78 m (RM 29). The general dip is between 3° and 4° to the south-east, this general trend agreeing with that mapped for the subcrop. In this part of the subcrop there is a branching system of subsidence features, largely of crater type and related to one-time brine abstraction along the Weaver valley (see also p. 52 and Figure 18). The main arm, marking a brine-run, trends eastwards close to the south side of the Moulton Fault from Marton Hole [618 674] to the Winsford Fault where the line is displaced to the south. Marton Hole, the most distant from the abstraction point at Newbridge of the subsidence points on the main arm has, nevertheless, been established for well over a century, and is the largest of the crater subsidences in the district. As stated on p. 42, it almost certainly owes its form to the existence of water-bearing glacial sands at the site.

Joseph Dickinson, H.M. Inspector of Mines, reported in 1882 (Dickinson, 1882, p.94) as follows. 'The farmer and the old people of the neighbourhood say that formerly it was but a small field pond, and that in 1863 the water suddenly disappeared from it, and that soon afterwards the bottom dropped perpendicularly down, but the bottom getting choked water filled the lower part again. In about three years afterwards the water again disappeared, and was followed in a day or two by another perpendicular subsidence, which increased the size of the hole to about 60 yards in circumference, the water coming again as before. In September 1870 the water again disappeared, and was followed by a further perpendicular subsidence, which shelved off, leaving the hole about 210 yards in circumference. On the morning of the 23rd December 1871, the water having left, it again sank perpendicularly to a depth of about 100 feet. The hole again became choked, the sides having shelved off, water stands, say 40 or 50 feet from the top, with tufty grass growing on the slopes above, and the circumference at the surface is increased to about 300 yards'. Marton Hole was actively subsiding when C. E. De Rance mapped the ground in 1892, and he noted subsidence scarps in the drift around the collapsing area. At the time of the 1956 survey the area of collapse was extending southwards, and there had been some sinking since the Ordnance Survey's 200-foot contour was drawn in 1911 to pass close to the depression.

Another circular crater subsidence, 'Bark House Hole' [6387 6872], now about 50 m wide and 16 m deep, formed in about 1876 in sand in a field on the north side of Pettypool Brook, 260 m east of Bark House. It is said to have subsided suddenly during the harvest.

This crater is close to the Winsford Fault, where the edge of the salt is displaced southwards, and the axis of the brine run, which must follow the salt, is likewise deflected. Its course along the Pettypool Brook valley towards an old abstraction point [6503 6893] at the now abandoned Newbridge salt works, is marked by a group of sink-holes [642 686], noted by De Rance in the 1892 survey and not now visible, and a deep circular pit [6425 6856] in the floor of the valley known locally as 'Tommy's Hole'.

Most of the subsidences on the main arm are of the conical or crater type, and formed rapidly by the funnelling of sand into subjacent cavities (see p. 42). An exception is a part of the Bogart Brook valley-bottom that has subsided in a broader downwarp to form a lake [632 682] through which the stream runs. From here, there extends southwards a system of subsidences characteristic of areas where the surface is of boulder clay (see p. 43). In the 1956 survey the southern tip of this branch had reached a point [635 669] 640 m north of St. John's Church, Over. De Rance noted that this group of subsidences was active enough in 1892 to produce subsidence scarps at the surface in the Gavel Green [6312 6784] and Old Brook House [6341 6809] areas, and buildings there were later destroyed. The Cheshire Lines Railway (now disused) crossed this southern branch of subsidence in the vicinity of Bradfordwood [6366 6761]. Between 1909 and 1956 the railway bank was raised 2.5 m in all, to maintain a level track across the brine run, and in 1956 alone it was raised 23 cm. Large ponds now occupy the main depressions.

The volume of all these depressions represents very large tonnages of salt removed as brine. Dickinson (1882, p.94) had no doubt that the cause of the Marton system of subsidences was brine abstraction at the Newbridge works, and he noted a succession of subsidences between the two places, which are some 3.6 km apart. These, together with those noted in the 1892 and 1956 surveys, are indicated on Figure 18.

A series of ill-drained depressions extends from Marton Hole southwards to a point [617 688] 280 m west of Chesterlane Farm. This may indicate a southern extension to the distal end of the main arm. The ground here is of boulder clay and the depressions are of the broad extended type.

The Over Borehole [6364 6587] was drilled in 1958 in a deep sand pit near the crest of a sand-cored ridge (Appendix 1, p.105 and Figure 14). The base of the drift is at 20.12 and 90.22 m of red and grey banded mudstone and silty mudstone with numerous gypsum veins, and 10.36 m of solution residues and collapsed mudstone, overlie salt with a 'wet' rockhead at 120.70 m (67.36 m below OD). By comparison with the highest salt beds in the Worleston Borehole [6585 5749], where the upper salt surface is intact because of its greater depth, it appears that about 17.5 m of salt have been removed by solution. In other respects the succession is normal for the Northwich Halite though the hole was abandoned at 305.71 m before the base of the Bottom Bed had been reached: this base is estimated to lie at 320.04 m below surface (266.7 m below OD). On the basis of the 'wet' rockhead this borehole is mapped as lying within the Northwich Halite subcrop.

Winsford No. 6 Borehole [6413 6512] was cored continously from the base of the drift at 32 m to the bottom of the hole at 226 m. The log (Appendix 1, p.119 and Figure 14) therefore, provides a useful standard for correlating the electric logs from the other four holes. Rockhead is at 158.5 m (97.7 m below OD), and overlain by about 4 m of collapsed strata and solution residue. Comparing the section with the Winsford No. 5 and Worleston boreholes it seems that not much more than 2 m of salt have been removed by solution, and the subcrop boundary line of the Northwich Halite on the dry rockhead side is drawn to pass close by the south side of this site.

Winsford No. 5 Borehole [6481 6453] was drilled open-hole to 170.7 m and cored to the bottom at 206 m. The upper part of the record is based on the electric logs (Appendix 1, p.108 and Figure 14). The top of the salt is drawn on the evidence of the gamma log at 166.42 m. Comparison with the Worleston Borehole indicates that there has been no significant solution at the salt top, which can be deemed dry. The line bounding the subcrop of the saliferous beds on its down-dip side is, therefore, drawn to pass close to the north of this site and well to the south of Winsford No. 1, where there has been appreciable solution of salt. The base of the Bottom Bed is estimated to lie at 384.66 m below surface (326.16 m below OD).

Winsford No. 1 Borehole [6451 6511] was drilled by rock-bit to 152.4 m, and cored to the bottom of the hole at 203.7 m. Gamma ray, high resolution density and neutron-neutron logs were made. The top of the salt is at 162.5 m (119.2 m below OD), and is overlain by what was described as a 'solution residue' though part of this may be collapse autobreccia (see p. 43). By comparison with the complete succession of the upper salt beds at the Worleston and Winsford No. 5 boreholes an estimated 12 m of salt has been dissolved. Because of the 'wet' rockhead the bore is mapped as being in the subcrop area of the Northwich Halite. The old Church Hill brine borehole of 1894 [6553 6503] lies about 140 m to the south-west, and signs of subsidence have been noted not far to the south-west of these bores. The gamma log of the measures above the salt between 146.3 and 156.97 m shows a rhythmic pattern characteristic of these beds. The pattern is found also in the other Winsford boreholes, but most importantly in Winsford No. 6 gamma log (see above), where it can be related to the lithological record from core examinations.

Two further isolated subsidence sites were noted well to the south of the Marton-Newbridge system, and on the downthrow side of the Ways Green Fault. One [6457 6689] is a saucer-shaped hollow elongated to the north in an otherwise featureless clay area. This site is close to the north or upthrow side of the line of the Bostock Fault (Figure 18). The other subsidence area is located 160 m due south of the Winsford No. 1 Borehole which stands in the southern angle between the Clive and Ways Green faults, and about 200 m from each. Wet rockhead was proved at Winsford No. 1 Borehole, and the old Church Hill brine-well site is nearby.

### Bowyers Waste and Oakmere

The aim of the Bowyers Waste Borehole [5686 6881] was to prove the nature of the heavily drift-covered ground east of the Overton Fault, where there were no solid exposures, but some indications of instability due to subsidence. The borehole proved a salt succession (Appendix 1, p.97 and Figure 18) with a 'wet' rockhead at 139.29 m (62.02 m below OD). The beds between 152.78 and 164.90 m, separating two thick salt beds, have the characteristic features of the Thirty-Foot Marl, including a thin salt bed between 161.72 and 164.03 m like those seen in a similar position in the Southley Common and Over boreholes (see Figure 14). The salt is overlain by a sequence of collapsed and autobrecciated mudstone 37.79 m thick with a few salt inclusions in the lower part. Above this, up to the base of the drift at 33.68 m, mainly red and grey banded mudstones and silty mudstones lie almost horizontally, but are broken and faulted at intervals, and heavily impregnated throughout with gypsum veinlets. The shattering is probably the result of a progressive lowering of the strata by piece-meal solution of the underlying salt, through a vertical distance of some 150 m. The gypsum, ubiquitous in finely disseminated form in the body of the rock, would tend to crystallise in the fine ramifying fissures caused by the collapse. Although the lower salt bed here is only 11.75 m thick it is correlated with the Bottom Bed of Winsford and Northwich. The lowest 10 m or so of salt and 15 m of the underlying mudstone with salt layers that are present in the nearby Oakmere Borehole (see below) are thought to be cut out in the hole by a fault, and the salt as proved rests sharply upon 0.31 m of pulverised mudstone. Using the Oakmere Borehole for comparison, the throw of the fault is estimated at about 27 m, for the beds below the fault plane carry salt inclusions, like those between 200 and 207 m in the Oakmere Borehole and in a similar position elsewhere in the saltfield.

It had long been suspected that salt was present beneath Oakmere. In his report on the Salt Districts, Dickinson (1882, p. 106) quotes H. E. Falk (who was seeking to prove that most solution of the salt took place naturally and was not due to brine pumping) as saying that Oakmere and an adjacent common were 'steeped in saline solution', and that there were unaccountable variations in the level of the lake. Following the discovery of salt at Bowyers Waste the Cheshire Brine Subsidence Compensation Board commissioned the Oakmere Borehole [5768 6780] on the eastern side of Oakmere to investigate alleged subsidences in that area (see Appendix 1, p. 104). 'The bore proved a thin bed of salt at 156.67 m, separated by about a metre of mudstone from a bed of salt below 23.43 m thick. This bed is confidently correlated with the Bottom Bed of the main saltfield for the following reasons. Its thickness compares well with other provings of this unit, particularly that at Southley Common (Appendix 1, p.107) which lies in a similar position with respect to the structural basin. Below the thick bed is an alternating sequence of mudstones and thin salt layers underlain in turn by mudstones with salt inclusions and veins, the whole being closely comparable with the sequence below the Bottom Bed at Winsford Mine and in several of the exploratory holes of ICI, including RM 9, RM 10 and RM 11. The main salt bed is overlain by 10.98 m of mudstone with only slight brecciation, and a thin salt bed near its base. This sequence agrees with the Thirty-Foot Marl section. All the salt above it has been removed by solution. In its place there is a 103 m sequence of collapsed mudstone, some parts totally disintegrated into autobreccia, others having retained stratification and carrying gypsum veinlets as in the Bowyers Waste Borehole. The autobrecciation is most severe in the lower part of the sequence, again as at Bowyers Waste.

In assessing the extent of the area underlain by the salt proved in these boreholes there arises the general difficulty in the Oakmere area, east of the East Delamere Fault, of distinguishing those tracts of peaty ill-drained land which are due to salt subsidence from those resulting from depositional irregularities in the Quaternary sands. The problem is not straightforward. For instance, three elongated peat patches [561 697, 563 691 and 566 685] trend roughly parallel to a strong Helsby Sandstone feature to the west. It might have been thought that the East Delamere Fault ran between these mosses and the rising ground were it not for a water well [5654 6861], sited within the southernmost of the three peat areas, which proved 16.61 m of peat and sandy drift on red sandstone to 25.60 m. It appears, therefore, that only Oakmere, the small peat moss east of the Bowyers Waste Borehole, and another [568 694] north-west of Delamere Lodge can be ascribed to salt solution with any certainty. The other mosses and ponds on the downthrow side of the Overton Fault and north of the Delamere Fault, including Linmer and Blakemere mosses, Hatchmere and Flaxmere, are not, on present evidence, considered to be underlain by saliferous beds.

### The northern limit of the saltfield west of Winsford

During the survey (1953–58) evidence of the precise position of the edge of the salt sub-crop between Oakmere and Moulton was inadequate, and the mapped line was therefore highly conjectural. The three main factors originally considered were as follows: Salt was proved at the Bowyers Waste Borehole where the top of it was undergoing solution and was overlain by collapsed mudstone. The Austins Springs Borehole [5952 6722] proved a thick succession of lower mudstone; the driller recorded nearly 7 m of 'broken red marl with gypsum' immediately beneath the drift, and this may represent the zone of collapsed ground fringing the salt subcrop where it is unfaulted (see p. 43). There was a series of subsidence features, apparently all part of one branching system, between Marton Hole and Bark House thence via the lower part of the Pettypool Brook valley south of Parkside Farm [6416 6892] towards the Weaver valley, where in the past there had been many brine abstraction points. A wide area from the Overton Fault to the Winnington Fault at Northwich contained no records of salt or subsidence, and this ground was presumed to be underlain by lower mudstone, a supposition strengthened by the Crabtree Green Borehole (Appendix 1, p.98) which proved 247 m of lower mudstone, resting on Tarporley Siltstones. To satisfy all the above evidence the line between the stable salt-free ground to the north and the proved salt occurrences and subsidences was drawn as a simple curve running south of Austins Springs, with a displacement at the Winsford Fault.

All the facts stated above are still valid, but since the survey further information, chiefly from an extension of the exploratory programme of Imperial Chemical Industries Ltd., has allowed some refinement of the map of the salt area. For instance, it is now plain that the Winnington Fault cannot exist as an important structure far south of Moulton. This gives more credence to a suggestion that had already been made on the basis of gravity measurements, (report by the Geophysical Prospecting Company to ICI in 1958) that a significant fault—the Moulton Fault—crossed the Weaver valley north of the Meadowbank Mine area, running through Parkside Farm [6416 6892] and the southern environs of Moulton throwing down to the south to form the local northern boundary of the saltfield. It is thought to continue west of the Ways Green Fault, trending WSW and passing to the north of the Marton-Bark House subsidences. A strip of ground on the upthrow side of the Parkside Fault for a kilometre or so west of the Ways Green Fault is considered to be underlain by a thin upthrown remnant of salt undergoing solution. The evidence for this is a crater subsidence [6332 6878] in sand on the south-west side of Pettypool Brook. The throw of the fault is hereabouts estimated at about 75 m. West of the Cassia Green to Marton road the fault forms the boundary of the saltfield, as far as a point [6000 6676] some 660 m SE of the Austins Springs Borehole. North-west of here the original conjectural line showing a normal unfaulted edge is retained, and the fault is discontinued for lack of evidence.

### Darnhall, Church Minshull, Worleston and Wardle

The solid rocks are obscured by thick glacial drift except at intervals along the bottoms of the Weaver valley and its tributaries. Most of the information is, therefore, from the fully-cored boreholes at Wardle and Worleston, and from Winsford No. Borehole which was made for planning purposes on behalf of the Cheshire County Council.

The Worleston Borehole [6585 5749] proved a 'dry' rockhead at 142.49 m (96.65 m below OD) beneath 127 m of undisturbed middle mudstone (Appendix 1, p.119 and Figure 14). Drilling was stopped at 226.82 m when the base of a halite unit at 226.67 m was correlated with a halite base in the Over Borehole [6364 6587] at 189.59 m.

The Wardle Borehole [6063 5713] (Appendix 1, p.107) proved 'dry' rockhead at 210 m (155.14 m below OD) beneath 195.37 m of middle mudstone. This depth of burial precludes any solution of the salt from present-day circulating ground-waters; nevertheless, there is a solution residue between 203.3 and 206.04 m. This residue is apparently the result of the local removal of the whole of the uppermost salt unit—that proved between 149.49 and 149.94 m in the Worleston Borehole. The underlying saliferous mudstone unit between the solution residue and the top of the salt at Wardle has remained intact. The solution residue is well-consolidated, the core being recovered intact with little tendency to disintegrate. It is probably the result of solution which took place not long after the deposition of the salt and its burial under mud and silt.

Winsford No. 3 Borehole was drilled uncored to 202.69 m and the record (Figure 14) is based on the electric logs. The hole was cased to 43.89 m. Throughout the uncased part below that level gamma radiation remained at a level indicative of mudstone; it is therefore, concluded that salt was not reached. The gamma record

below about 175 m does, however, show distinct similarity to that of Winsford No, 6 Borehole between 123 m and the top of the salt at about 159 m. The upper part of the fourfold rhythmic pattern in the middle mudstone overlying the salt is present near the bottom of the hole, and a minor peak on this oscillating sequence at 202.08 m is correlated with a similar feature at 149.96 m in Winsford No. 6 Borehole and at 156.36 m in Winsford No. 5. On this basis rockhead, presumably 'dry' lies about 8.81 m below the bottom of the hole, and the base of the Bottom Bed 218.24 m lower at 429.74 m. The Winsford Fault must pass between this borehole and Winsford No. 2, and the throw of the fault may be upwards of 180 m. The 'dry' rockhead inferred here agrees with the distribution of surface subsidence features, which are not detectable west of the Winsford Fault hereabouts.

### East of the Winsford Fault

A belt of country some 2 km wide, extending from north-west of Winsford, eastwards between Winsford and Moulton, and thence through Bostock parish to beyond the river Dane, was intensively explored by Imperial Chemical Industries Ltd., between 1956 and 1977. Most of the boreholes were uncored, the sections being interpreted from electric logs; enough cored holes were, however, made to provide adequate lithological references for the electric logs. Before this programme, precise information on the saliferous beds at Winsford was restricted to that obtained from the two shafts at Meadowbank Mine (see Appendix 1, p.102 and Figure 14). A list of 71 brine and rock salt shafts along the Weaver valley at Winsford between the north end of Bottom Flash and Newbridge was published by Calvert (1915, pp. 260–263). A site map of these shafts has been kindly provided by the Cheshire Brine Subsidence Compensation Board (Figure 19). Apart from their value in locating the subcrop of the Northwich Halite, these records provide little hard geological evidence. Few traces of these workings were visible at the time of the resurvey.

South of Winsford the Weaver valley still bears the scars of subsidence from a ramifying system of brine-runs that supplied the brine industry of Winsford for more than a century. The salt works lay on both sides of the river from a short distance south of Winsford Bridge to Newbridge, a distance of about 3 km. In 1881 there were 28 brine pits at Winsford, with 556 evaporating-pans, giving an annual production of salt from brine of 1 million tons (Dickinson 1882, p. 87). The overall subsidence pattern around the town suggests that only pits in the vicinity of Newbridge drew on the brine from the direction of Marton Hole, and that the main concentration of pits south of Newbridge drew brine from south of the town. The greatest effect is seen along the floor of the Weaver valley, where the river has widened into three lakes or 'flashes' following the differential lowering of the valley floor. In 1882, Dickinson (p. 93) described the development of the subsidence as follows:

'At Winsford for about 2½ miles in length by one in breadth the surface is visibly on the move. The advancing front up the river Weaver is about 2½ miles from the nearest brine shaft, and 3¼ miles from the furthest at Winsford. At Wimboldsley, near the front, the canal has seriously sunk, and the over-bridge of the London and North Western Railway, between Winsford and Minshull Vernon stations, on the line from Crewe to Warrington, has been broken and had to be rebuilt on girders so as to enable it to be lifted as may be required. At the Winsford end the subsidence is backing down into the town. In the intermediate parts, in the valley of the Weaver, two large meres called flashes have formed, and the river flows through them, whilst on the high ground the land is being torn into furrows and holes, and the subsidence is rapidly extending. Where the subsidence and slips are in fields or woods the damage is confined chiefly to the vegetation and timber, with some risk to cattle and persons, but when buildings are in the way there is rough work. In the town the church has had to be taken down, and many buildings may generally be seen shrunk and cracked, notwithstanding the frequent rebuilding. New buildings are being placed upon strong wooden frames, so that they may be lifted when required, and others have iron binders preparatory for the landslip when it comes'.

By 1956, subsidence in the Weaver valley had not advanced much farther south than the position of the 'front' noted by Dickinson, although brine abstraction continued at a high rate at Winsford for a further half-century. The inference is that it had reached some limit, and it is concluded that the subcrop or 'wet' rockhead boundary passes close to the limit of subsidences hereabouts. This is borne out by the results of Winsford No. 2 Borehole, together with a re-appraisal of three old records, those of the Stanthorne, Wimboldsley and Brook Farm boreholes.

Winsford No. 2 Borehole [6604 6391] was drilled uncored to 153.9 m, and the record (Figure 14) is based on electric logs. 'Wet' rockhead is at 118.3 m (75.63 m below OD). The salt section derived from the gamma log correlates well with the Over Borehole, the bottom of Winsford No. 2 corresponding with the base of a mudstone bed at 235.66 m at Over. The distinctive fourfold rhythmic gamma log pattern found immediately above the Northwich Halite is well-marked in the gamma record between 68.28 and 86.87 m. As the salt rockhead here is at 118.3 m, it is apparent that 31.09 m of rock resembling mudstone on the gamma log intervene between the base of this pattern (which also marks the base of the middle mudstone) and the salt top, and this sequence is interpreted as a solution residue. Comparison with the section of higher salt beds in the Worleston Borehole where the full succession is present, shows that salt has been totally dissolved from the upper 94 m of the Northwich Halite. At Worleston the mudstones in this sequence add up to 27.74 m. This leaves 3.35 m to be accounted for in the 31.09 m of residue at Winsford No. 2. Thus, from a total of 66.26 m of salt, there has been derived 3.35 m, or about 5 per cent of included muddy material. This is a figure widely accepted as an approximate average for impurity in the salt in Cheshire. The purpose of the above calculation is to confirm that the correlation of this hole is correct, by showing that the rockhead is overlain by a solution residue of the precise thickness required to support it. It then follows that the Winsford Fault must pass to the west of this hole, leaving the borehole on its upthrow side where the 'wet' rockhead conditions prevail. For structural purposes the base of the Bottom Bed here is estimated at 238.30 m below surface or 195.63 m below OD.

The site of the Stanthorne Borehole is given by Sherlock (1921, p. 34) as 'two miles ESE of Winsford Bridge and on the south side of the canal'. A capped casing tube protruding from the ground near the junction of four roads [6882 6495], some 350 m from Sherlock's point, is now deemed to mark the site of this bore (Figure 16). It had originally been assumed that 10.97 m of mudstone overlying 24.99 m of salt in the lower part of this section respectively represented the Thirty-Foot Marl and the Bottom Bed of the Northwich succession, because in Sherlock's time the full succession was not known. The revised correlation is as shown in Figure 14. Assuming the corrections of this correlation, an estimated 56 m of strata have been removed by solution, and this confirms the mapped position of the bore in relation to the subcrop. The ground immediately above the salt was described by the driller as 'shaggy marl with brine-run'. 'Wet' rockhead was at 67.67 m (25 m below OD).

The Wimboldsley Borehole is marked by its capped and protruding casing close to the site given by Sherlock (1921, p. 32). As in the case of the Stanthorne Borehole, and for the same reason, he assumed that a mudstone bed 11.28 m thick overlying 18.29 m of salt represented the Thirty-Foot Marl and Bottom Bed of Northwich. The revised correlation is shown in Figure 14. The similarity of this and the Stanthorne sections implies that the two bores are roughly aligned along the strike, and this agrees with a dominant trend in the linear subsidences on the flanks of the Weaver valley

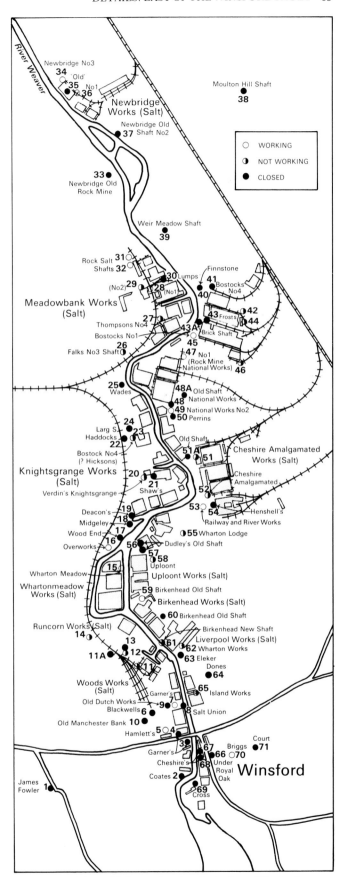

**Figure 19**  Map of brine and rock salt shafts at Winsford c.1910. For details see Calvert, 1915, pp.260–263

(see Figure 16). 'Wet' rockhead was reached at 80.16 m (36.27 m below OD).

The Brook Farm Borehole near Minshull Vernon was correctly sited by Sherlock, who incorrectly named the locality Brook House. The site [6797 6135] is marked by its protruding casing tube capped by a cover bearing the name 'F. Timmins', and the Ordnance survey 1:10 560 map refers to it as 'Shaft (Rock Salt)'. The section correlates bed for bed with the upper part of the saliferous beds in the Worleston Borehole. The full succession is, therefore, present and the rockhead, reached at 103.02 m (66.9 m below OD) is presumed to be 'dry'. This bore gives further support for the mapped position of the down-dip subcrop boundary line of the Northwich Halite between Wimboldsley and the Winsford Fault.

South of the Bottom Flash several of the subsidence depressions on the valley sides show a marked ESE-WSW linearity which may be the surface manifestation of the strike of the beds (see p.42). The brine-runs under these linear depressions apparently fed a main, possibly rather broad, corridor of brine movement running northwards along the centre of the Weaver valley towards Winsford. The linear subsidences need to be distinguished carefully in the field from surface drainage channels. Their centre lines do not show a systematic fall when traced down a valley side, and on flat land they may connect a line of ponds or ill-drained patches, as in the example [665 638] north of Weaverwood Farm. Where they cross the floor of the Weaver valley they have given rise to several minor flashes. Drainage channel and linear subsidence can be seen almost side by side on the hillside between Weaver Dairy House [6615 6476] and Weaverhall Lane. A stream draining the sands which crop out in Badger Wood [660 642] flows north-east to join the Weaver between two flashes. About 200 m east of Badger Wood is the beginning of a marked linear depression which trends ENE through several ill-drained patches of ground. Its junction with the Weaver is marked by the development of a flash (unnamed) of some 4.2 hectares. The line of this depression is continued on the east side of the Weaver in a curving linear subsidence extending towards Clive Green.

On the east side of the valley, in addition to the depression in the Clive Green area noted above, there is a flat-bottomed subsidence feature running from about 200 m SW of Clive House [6713 6537] towards the Weaver; two converging subsidence features south of Clive Farm [6648 6545] appear to be in linear continuity with a subsidence on the west side of the river east of Weaver Dairy House [6615 6475].

At the top (southern) end of the Bottom Flash the style of subsidence changes somewhat. The main effects from here northwards are more closely confined to the Weaver valley, and it would seem that the movement of brine is via a system of sub-parallel brine-ways a few hundreds of metres wide on the 'wet' rockhead beneath the valley floor. At the time of survey (1953) subsidence scarps were particularly conspicuous on the west side of the valley in the fields leading down to the shore of the flash, and some were still active in 1979 some 15 years after the abstraction of natural brine had ceased at Winsford.

Dickinson (1882 p.95) summarised damage to property in the town of Winsford. Little of the evidence survives, because demolitions, repairs and the infilling of ground went on throughout the period of exploitation of brine. It is clear, however, that the salt extends beneath the whole area, and some of the numerous brine shafts and rock-salt mines by which the deposit was exploited for at least two and a half centuries are shown in Figure 19.      BJT

## MIDDLE MUDSTONE DIVISION (formerly 'Middle Keuper Marl')

This formation comprises a rhythmic sedimentary sequence of mainly red, reddish grey and greenish grey mudstone and siltstone, alternating between a facies in which the components are sharply separated in thin laminae, and a blocky or unstructured facies with the components intimately mixed. Anhydrite occurs as nodules; dolomite occurs widely, cementing some siltstone laminae into hard platy rock; gypsum occurs as displacive veins, and halite is sporadically present as veins or other inclusions, chiefly near the base.

The outcrop of the formation is largely drift-covered, and no borehole within the district has penetrated the entire sequence. Composite sections and field evidence suggest a total thickness of at least 320 m; the Wilkesley Borehole in the Nantwich (122) district to the south proved a thickness of 327 m (Poole and Whiteman, 1966, pp.125–129). However, a few kilometres to the east of this district, and east of the King Street Fault, the Byley Borehole proved the lowermost 427 m of the formation, and mapping suggests the presence of a further substantial thickness (Evans and others, 1968, p. 135). Recent exploration by Imperial Chemical Industries Ltd., tends to substantiate the higher estimate for the formation. As has been pointed out elsewhere (p.34) there is some evidence that the Cheshire Graben was to some extent synsedimentary and, therefore, thicknesses within it may be greater than those of equivalent strata on its flanks. This could account for the discrepancy between thicknesses to the east and west of the King Street Fault.

The middle mudstone division occupies the ground between the subcrop of the Northwich Halite and the postulated subcrop of the Wilkesley Halite around Wardle, Wettenhall and Darnhall; it also occurs east of the Winsford Fault from Church Minshull to Worleston.

The formation is also present in the subcrop area of the Northwich Halite. Figure 15 explains the relationships. Wherever the formation occurs in the salt-crop areas it has been displaced vertically to a greater or lesser degree by the removal of salt below. Near the down-dip edge of the subcrop the displacement is small; near the up-dip edge the mudstone may have moved through hundreds of metres. Cored boreholes through the mudstones above 'wet' rockhead, such as the Southley Common, Over and Bowyer's Waste boreholes (logs given in Appendix 1) show that this subsidence can take place without gross disruption of the collapsed beds. Beneath the river Weaver, from the Ash Brook confluence northwards to the Moulton Fault north of Winsford, the middle mudstones are draped over the edges of progressively lower salt beds. In the river banks and tributary gullies these displaced mudstones are seen with dips variable both in angle and direction. Figure 15 also shows that the mudstone stratigraphically above saliferous strata may extend, at outcrop, beyond the limit of the salt on the up-dip side, taking up a mapped position apparently below the saliferous beds. These field relationships must be borne in mind by collectors of specimens for stratigraphical purposes from the middle mudstone division.

## Details

*Darnhall, Wettenhall Brook and Ash Brook*

There are no exposures west of the Winsford Fault in the area covered by the Marton-Crewe esker ridge north of Ash Brook. Winsford Nos. 3, 5 and 6 boreholes, all made for the Cheshire

County Council (Appendix 1, pp.108–9), proved beds of the middle mudstone division in areas where they are not affected by solution of the salt below. A thin solution residue is recorded near the base of the formation in Winsford No. 6 Borehole, [6413 6512] but less than one metre (if any) of the underlying salt appears to have been removed, as can be seen by comparing the sequence with the full succession at the Winsford No. 5 and Worleston boreholes. The solution residue may result from the solution of halite inclusions within saliferous mudstones proved at the base of the formation at the Worleston Borehole.

The cyclic character of the middle mudstone division has been described in detail by Arthurton (1980). One of the sequences that he described is from Winsford No. 6 Borehole where 26 rhythmic units occupy about 120 m of strata between the base of the drift and the thin solution residue above the Northwich Halite. Each unit comprises finely banded grey and red mudstones and micaceous siltstones, dolomitic in places, at the base, and unstructured blocky red silty mudstone at the top. The average thickness of the cyclic units is about 5 m.

Winsford No. 6 is the only borehole in the district which provides a direct comparison between lithological and geophysical logs. The unpublished data show the close relationship of the two logs in 7 complete rhythmic units between the bottom of the steel casing at 123 m and the base of the formation at 158.5 m. The gamma peaks coincide with the red blocky layers, and presumably reflect the abundance of combined potassium in illite and the micas. This suggests that the banded facies contains a higher proportion of non-radioactive constituents such as detrital quartz silt, than the blocky facies.

The characteristic pattern of four gamma peaks, diminishing downwards, between 145 m and the top of the Northwich Halite in Winsford No. 6, has been recognized in the logs of other boreholes, notably in the uncored Winsford No. 2, where the fourfold pattern greatly assisted the correlation of the salt strata below (see p.53). At Winsford No. 1 Borehole [6541 6511] some 150 m of this division were drilled but only the lowest 10 m were cored. The gamma record between 126 m and the top of the Northwich Halite at 162.6 m shows a clear correlation with the uncased part of this division in Winsford No. 6. Winsford No. 3 Borehole [6556 6400] went through about 190 m of the division ending at 202.69 m, an estimated 9 m short of its base on the basis of gamma log correlation with Winsford Nos. 1 and 6 boreholes.

Beds exposed in Ash Brook between a point 30 m upstream of Darnhall Bridge [6366 6315] and the confluence with Wettenhall Brook [6409 6244] are of red blocky facies, flat lying or very gently dipping south-east. Other exposures lie in Wettenhall Brook and its tributaries, from the environs of Wettenhall village to the confluence with Ash Brook, and thence down Ash Brook to the point [6592 6258] where the Winsford Fault is thought to cross the stream. Throughout most of this 3.5 km stream section the strata are either horizontal or dip very gently south-east. A laminated facies comprising finely banded red and grey mudstone and siltstone, and a blocky facies of structureless red or red grey mottled mudstone are both present, but exposures are discontinuous and the compilation of a continuous sequence from them would be difficult if not impossible. Close to the mapped position of the Winsford Fault there is local disturbance of the bedding.

## Wardle, Haughton and Barbridge

The Wardle Borehole [6063 5713] proved 195.37 m of the middle mudstone division between the base of the drift at 14.63 and 210 m, the top of the Northwich Halite. The upper part of the hole provided cores stratigraphically higher than any obtained elsewhere in the district, although equivalent strata were drilled in the uncored Winsford No. 3 Borehole, and they may also be present

amongst the exposures in Ash Brook and the Haughton Hall–Barbridge stream (see below). The beds are alternations of red and grey banded and red blocky facies, as described above in the Winsford No. 6 Borehole, extending upwards from the top of the Northwich Halite to about 57 m below surface, a thickness of 153 m. The uppermost 37 m are generally structureless red-brown mudstone with occasional greenish grey mottlings. Samples of mudstone from this borehole were processed and examined for palynological residues. The only productive specimen was from 183.2 m, immediately above a prominent siltstone bed. Preservation was poor, the assemblage being heavily carbonised. One species of relatively restricted range suggested a Middle Triassic (late Anisian to early Ladinian) age for the bed. A full list is given on p.34.

The Southley Common Borehole [5816 5934] also proved the lower part of this division. The site in the subcrop area of the Northwich Halite and, although there are now few signs of surface subsidence in the vicinity at least 91 m of the underlying salt has been removed, and the middle mudstones let down by the same amount. There is no wholesale distortion of the strata, but some brecciation is recorded at intervals.

An unnamed brook which flows eastwards from near Haughton Hall to Barbridge has cut through boulder clay to expose beds in this division at intervals from a point [5926 5655] about 300 m east of Haughton Hall Farm to a point [6093 5654] 350 m WNW of Bullsgreen Farm. There is also an isolated exposure [6145 5644] 160 m east of Bullsgreen Farm. The spread of these exposures, with an easterly dip varying between 22° at the western exposure and 5° to 12° elsewhere, is about 2 km, and the thickness of the beds involved would be about 348 m if there were no faults. No idea of the detailed succession can be built up from these sporadic exposures, which include unstructured red or grey mudstone, and banded red and grey mudstone and siltstone.

## East of the Winsford Fault; (Church Minshull and Worleston)

The river Weaver, from where it enters the district in the south, to near its confluence with Ash Brook, is underlain by the middle mudstone division (Figure 16). To the north lies the subcrop of the Northwich Halite where the salt has undergone solution. The line between the two is highly conjectural and probably oversimplified. In the stable area south of the salt subcrop sporadic river bank exposures occur between a point [6723 6297], just west of a footbridge over the river, upstream to the road bridge over the river at Church Minshull. The exposures are generally of short sections of strata, only exceptionally more than 2 m in thickness, including grey and red mudstone both laminated and unstructured. The beds are either flat-lying or with variable dips up to 5°.

In Ash Brook, between 240 and 700 m east of Ashbrook Bridge [6563 6255], there is a change of dip direction to north-west and an increase to 10°. The Winsford Fault is drawn to pass close to the most westerly of the higher dips.

South of the road bridge at Church Minshull, as far as the western point of a river bend near Wades Green, scattered fragments of red or grey mudstone in the river banks indicate that bedrock is not far below surface. For the next 3.5 km upstream there is only a single exposure [6636 5827] of 0.5 m of micaceous grey and red mudstone; otherwise the river is in boulder clay. About 550 m upstream from the Leighton Brook confluence, 1 m of red structureless mudstone on 0.3 m of grey flaggy siltstone dips northeast at 3°. Both Wistaston Brook and the river Weaver above their confluence near the southern edge of the district have cut through boulder clay to flat lying blocky grey and red mudstone. The Worleston Borehole [6585 5749] proved the lowest 127 m of the middle mudstone division, consisting of alternations of red and grey laminated and red blocky facies. The junction with the saliferous beds below is a normal one, there having been no solution of salt. As in the Wardle and Winsford No. 6 boreholes the laminated beds are more arenaceous in the lowest 35 m of the sequence than higher up.

*Spurstow-Alpraham*

East of the East Delamere Fault in the Spurstow-Alpraham area the Mercia Mudstone has been encountered in scattered boreholes and exposures in the banks of the river Gowy and the Shropshire Union Canal.

At Spurstow Hall [569 565] a drilled well pierced 'Keuper Marl' to a depth of 82.3 m below 10 m of sandy drift. No water was obtained from the 'marl'.

About 900 m SSE of Spurstow Hall poor exposures of red and grey mudstone, dipping east at 20 to 25°, are seen under drift in the banks of an unnamed stream which runs 450 m west of Peartree Farm [577 556]. The stream runs parallel with the strike of the beds, and the exposures, spread over about 200 m, probably amount to no more than a metre or two of strata.

In the banks of the Gowy 27 m south of Tilstone Mill [567 593] at Tilstone Bank, about half a metre of red blocky mudstone is exposed, dipping west 43° north at 4°. Some 275 m east of this exposure, in the north bank of the Shropshire Union Canal, less than a metre of red and grey shaly mudstone was noted, dipping west 10° north at 3°.

# WILKESLEY HALITE

In the districts to the east and south, the Wilkesley Halite—named from the Wilkesley Borehole, where it was first completely proved—is widely preserved. Within the Chester district it is believed to occur beneath some 6 km² in the extreme south-east near Barbridge. The limits of the outcrop are, however, speculative, although boreholes at Hurleston Reservoir [624 554] in the district to the south proved salt solution collapse breccias of red mudstone.

# CHAPTER 5

# Structure

The Chester district lies on the western flank of the Cheshire Basin, a wide post-Carboniferous pull-apart basin, which, together with the Worcester Basin to the south, provided the main northward route for Permian and Triassic arenaceous sediments into the Irish Sea Basin. Figure 20 shows the structure of this district in the setting of the surrounding areas.

Using gravity data and published Geological Survey maps, White (1949) found that the main part of the Cheshire Basin was a deep trough bounded by parallel faults. He identified the main bounding faults as the Red Rock Fault on the east and the Delamere faults on the west, these corresponding to opposed belts of steep gravity gradients. The part of the Cheshire Basin lying west of the Delamere faults, including the western half of the Chester district, he considered to belong structurally to the adjacent massif of Wales and he called it the 'Chester Platform'. The main bounding faults of the Cheshire Basin continued to grow throughout Permian and Triassic times with the results that the Permo-Triassic sediments thicken appreciably towards the axis of the basin.

The BGS 1:250 000 Bouguer anomaly map (1977) indicates the clear differences in tectonic style of the two areas. Between the boundary faults there is a pronounced linear gravity low at − 7 to − 9 mgal; this passes through Nantwich, Crewe, Holmes Chapel and Alderley Edge and lies mainly to the east of the Chester district. Within the Cheshire Basin the isogals are parallel to the main bounding faults, but west of it this is not so, and the chief feature is due to the Carboniferous inlier around Milton Green where there is a pronounced high of + 20 mgal.

White (1949, p.353) considered that the western bounding faults were of insufficient magnitude at surface to account for the intensity of the observed gravity anomaly, and he concluded that these faults were initiated in pre-Permian times. The more recent understanding of the development of growth faults at the margins of pull-apart basins suggests that these western faults are growth faults; if so, the gravity anomaly may be due wholly to Permian and Triassic sediments.

Many previous accounts of the structures of the Cheshire Basin have, however, related the main episode of faulting to the mid-Tertiary 'Alpine Orogeny' (e.g. Taylor and others, 1963; Poole and Whiteman, 1966; Evans and others, 1968), but the views that the rift-bounding faults were at least partly contemporaneous with deposition have been expressed frequently. Strahan (1905) quoted Gibson's opinion that the boundary fault of the North Staffordshire Coalfield (Red Rock Fault) had a 900-yard throw in the Carboniferous but much less in the Trias, with the implication that the fault developed progressively over a long period. Hickling (1918a, b) first suggested that changes in thickness of the Permian Collyhurst Sandstone at Manchester, around the northern rim of the Cheshire Basin, were due to penecontemporaneous faulting. Audley-Charles (1970, pp.49 – 89) considered that the Worcester, Cheshire and Irish Sea grabens subsided at intervals through the deposition of the Trias, and more or less continuously during the deposition of most of the Mercia Mudstone Group. In the light of recent research it seems most likely that the whole rift system, from the Worcester Graben northwards to the Manx-Furness Basin in the eastern Irish Sea, was part of the tectonic activity that culminated in the opening of the north Atlantic. The rift system would then constitute a 'failed arm' of the north Atlantic, and its progressive development controlled sedimentation throughout Permian and Triassic times (Kent, 1975; Owen, 1976). The largest regional episodes of vertical movement were probably in the Permian and Triassic and immediately pre-Albian; mid-Tertiary times being one of the less active periods.

Seismic surveys (Kent, 1975; Gale and others, 1984) and boreholes (Colter and Barr, 1975) demonstrate the presence of up to 4000 m of Permo-Triassic sediments in the deepest part of the Cheshire Basin north-west of Crewe, and over 3000 m on the eastern side of the Chester district. The limited information in this district regarding the thicknesses of the Permian and Triassic formations shows that these thin westwards across the bounding faults of the Cheshire Basin. In particular the Lower Mottled Sandstone decreases from an estimated 600 m in the east to only 200 m at outcrop in the south-west.

From all available evidence it seems likely that there was a Permian phase of crustal tension which caused a wide and shallow downwarp to form between the Welsh uplands and the Pennine area, and normal faults to develop across this basin. There was then a second phase when most of the subsidence took place within the confines of the Cheshire Basin, in which great thicknesses of sediments accumulated.

White (1949) noted that an east – west zone of steep gravity gradients about 8 km south of Chester 'lines up well with an ancient surface fracture, the Llanelidan Fault which stems from the Bala Fault'. This zone passes about 2 km north of Rosset in the Flint (108) district to the west (Figure 20) and crosses the boundary of this district 3 km west of Aldford passing to the north of the Carboniferous inliers. The abrupt northward fall in gravity values across this line seems to imply an abrupt northward thickening of the basal Permo-Triassic sandstones, although no east – west faults have been recognised at surface. The Llanelidan Fault had an appreciably northerly downthrow in Carboniferous times (the southerly downthrow north of Wrexham is the result of the interaction of folding and faulting), and the northward thickening of the later rocks is considered to represent the filling of a pre-Permian depression bounded by a fault scarp along the Llanelidan Fault.

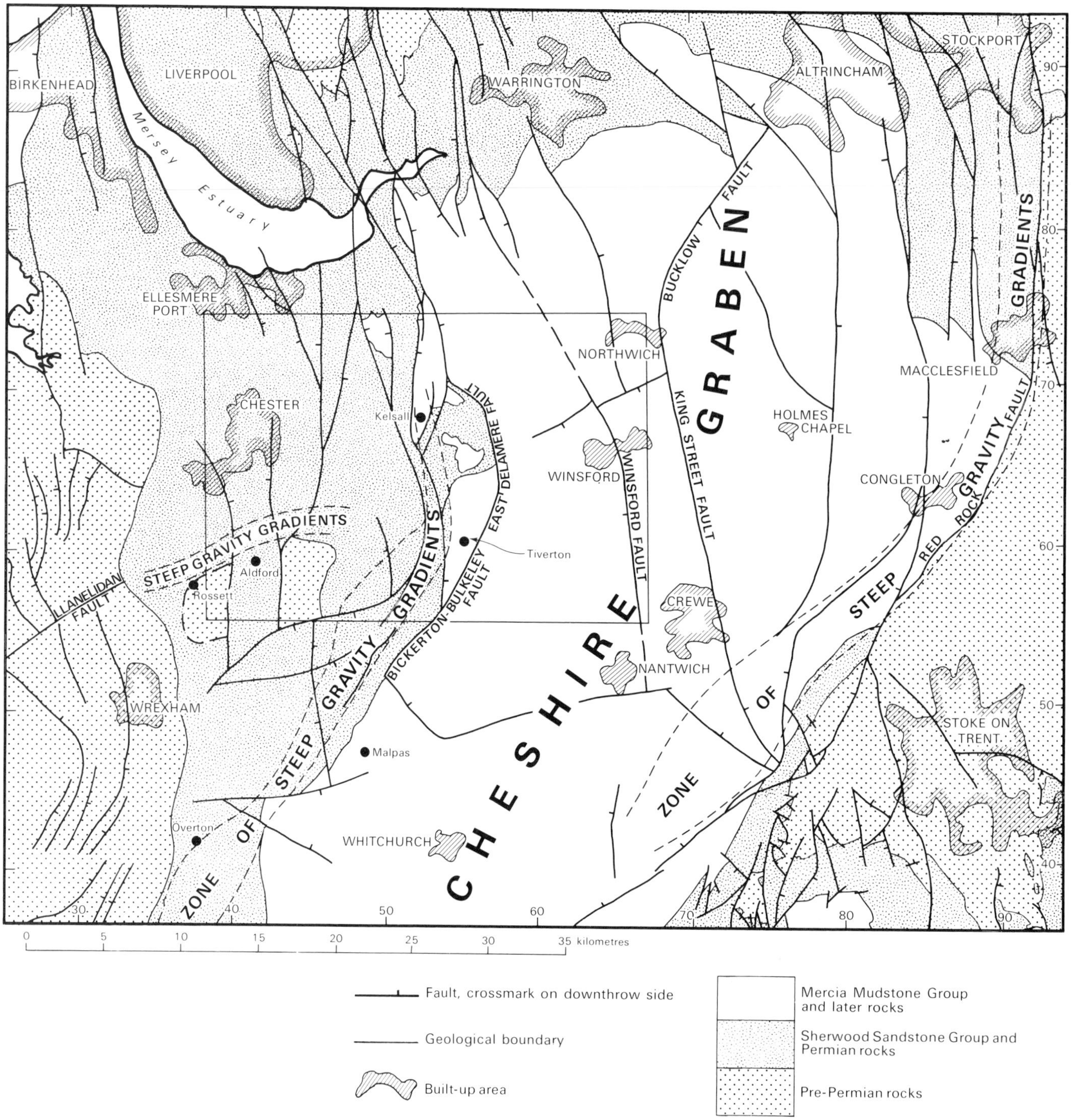

**Figure 20**   The Chester district in its structural setting

## FAULTS

Within the Chester district the regional trend of the main faults of the Cheshire Basin swings from NE–SW in the south to NNW–SSE in the north (Figure 20). Away from the mid-Cheshire range of hills the extensive drift cover locally makes the location of even the most important faults difficult to the point of conjecture from surface mapping alone. The positions of a number of the larger faults have been inferred from borehole and geophysical data

unavailable when the survey was carried out. Inevitably they differ from those shown in the earlier edition of this sheet.

On the mid-Cheshire hills, where exposures are better, there are several minor short faults, some of which throwdown to the west and so are antithetic to the main bounding western faults of the Cheshire Basin. Here, and at Winsford, where borehole evidence is plentiful, other minor faults at high angles to the main set are presumed to be contemporaneous with the latter, relieving local stresses within the main fault slices (see Figure 18). Tucker and Tucker

(1981, p. 495) have described linear patterns of fissures in the halite at Meadowbank Mine which suggest that tectonic stresses were acting on the salt beds while deposition was proceeding. There is also evidence in the detailed sections based on the boreholes of Imperial Chemical Industries Ltd., that these stresses may in the early stages have been relieved by monoclinal folds, the folding giving place to faulting as movement continued. For example, the un-named fault near the north end of Section C, Figure 21, is one of a pair of en-echelon fractures, along the general line of which the displacement is partly taken up by monoclinal folding.

## Details

### West of the Clotton Fault

The discovery of Erbistock Formation at outcrop by the Trevalyn Borehole (Figures 2 and 3) suggests that there is a N–S fault at Trefalun Meadows with an easterly downthrow opposed to the Edgerley Fault. The Edgerley Fault throws eastward-dipping Pebble Beds against the Upper Coal Measures of the Milton Green Inlier. Its continuation northwards to Chester is conjectural, as is the southern section of its branch, the Waverton Fault. The latter is clearer farther north, where it throws Pebble Beds against Lower Mottled Sandstone to the west of Barrow Hill and Dunham-on-the-Hill. The only other significant fault in this area appears to be a downthrow east with a NNE trend; this brings Upper Mottled Sandstone and the upper part of the Pebble Beds against the eastern flank of the conglomeratic sandstones of Barrow Hill and Dunham-on-the-Hill.

### Between the Clotton and East Delamere faults

This tract includes the main bounding growth faults on the west side of the Cheshire Basin, and encloses all the outcrop of the Tarporley Siltstones and Helsby Sandstone. The most easterly of the faults is the Clotton Fault which has a throw of about 300 m. It continues southwards into the Nantwich (122) district as the Holt-Coddington Fault (Poole and Whiteman, 1966, p. 53) and, to the north, joins the Frodsham Fault of the Runcorn (97) district (Figure 20). The Peckforton Fault trends roughly parallel to the Clotton Fault in the south of the district where it has a throw of about 200 m, but its throw decreases north of Kelsall and it is replaced en echelon by a minor fault at Crossley Hospital. It joins the East Delamere Fault at Peckforton and continues southwards with a throw of 800 m into the Nantwich (122) district as the Bickerton-Bulkeley Fault (Poole and Whiteman, 1966, p. 54). The third and most easterly of the growth faults is the East Delamere Fault which forms the western limit of most of the Mercia Mudstone outcrop. The presence of this fault north of Delamere, as shown on the previous edition of the map, has not been confirmed, and it is now thought to turn westwards to join the Overton Fault of the Runcorn (97) district. The maximum throw of the East Delamere Fault is 700 m where it brings Helsby Sandstone against Northwich Halite. Just beyond the northern edge of the district the Overton Fault is jointed by the en-echelon continuation of the Peckforton Fault, and the two have a combined throw of 150 m.

Within this area the Helsby Sandstone, the Tarporley Siltstones and the immediately associated beds are gently tilted to the south-east or, as in Delamere Forest, are flat-lying; some minor folding is described below. The sequence is commonly repeated by antithetic westward-throwing faults. Two of these west of the Primrose Hill Fault have throws in excess of 150m. In the better exposed areas a few E–W faults with small throws have been mapped, for example in the north at Simmond's Hill and in the south at Stanner Nab.

### East of the East Delamere Fault

The present structural interpretation of this largely drift-covered area relies heavily on geophysical and borehole data.

The Winsford Fault is established by the Over Borehole [6364 6587] which proved almost a full succession of the Northwich Halite, whereas along strike to the east in the Weaver valley there is a broad belt of unstable ground, clearly much lower in the Northwich Halite subcrop, which extends as far south as Weaverwood and Top Flash (see Figure 18). The fault necessary between these two areas must run west of all the subsidences in the Weaver valley; its line takes into consideration borehole data from the Imperial Chemical Industries Ltd., and the Cheshire County Council (Figure 18). The continuation of this fault north of Vale Royal is largely conjectural, but it probably joins with either the Preston Brook or the Keckwick faults of the Runcorn (97) district. The throw of the Winsford Fault is of the order of 120 to 140 m in north-west Winsford, variation being due in some measure to the effect of minor faults which meet or cross the main fracture. Farther south, between Winsford Nos. 2 and 3 boreholes, it increases to 200 m.

The Winnington Fault has been traced for 13 km in the Runcorn (97) district. From the western end of Budworth Mere southwards it separates the Northwich Halite on the east from the lower mudstone division on the west (Figure 17). In Northwich its course is indicated by borehole and shaft sections, but south of Davenham its course is obscure and there is no evidence for it south of Moulton. It may well lose much of its throw on meeting the Moulton Fault, and it dies out where it meets a minor fault at right angles to it [6627 6892], about 1.5 km WNW of Bostock Hall (Figure 18).

The presence of the Moulton Fault resolves a problem in determining the northern edge of the saltfield north and north-west of Winsford. The depth and volume of the subsidences between Marton Hole [618 674] and Newbridge (see p. 50) suggested that there was a substantial thickness of salt present, local removal of which by solution had produced large cavities; yet immediately to the north in the Whitegate and Vale Royal areas there were no signs of salt or subsidence. Boreholes made for Imperial Chemical Industries Ltd., now strongly suggest that at least as far west as Marton Hole, there is an ENE-fault throwing south and trending parallel to other minor faults proved in the Winsford area. West of the Winsford Fault (Figure 21) the throw of the Moulton Fault may be as much as 70 m. East of the Winsford Fault, the salt passes northwards into a zone of brecciated marl which appears to be the result of solution close to the fault. In the vicinity of the river Weaver the fault passes between boreholes RM 8 [6505 6918], which proved salt, and RM 6 [6479 6919], which proved beds resembling the Tarporley Siltstones at between 221 and 255 m depth. These beds may represent the higher 'Waterstones' phase proved in the Marston and Plumley boreholes (see p. 37) or they may indeed be the Tarporley Siltstones. In either event, RM 6 must lie on the upthrow side of the fault, the throw of which may well exceed 100 m at this point. The Moulton Fault may lose much of its throw or even terminate against the Winnington Fault near Moulton village.

The Bostock Fault is known only from ICI's exploratory bores. It trends ENE, like the Moulton and Church Hill faults, and throws down south (Figure 18). Between the King Street Fault and Bostock Hall the displacement is probably of the order of 75 m; this appears to diminish westwards to about 60 m where the fault-line crosses the railway [6613 6774].

The Church Hill Fault, also known only from boreholes sunk for ICI and the Cheshire County Council, trends ENE and throws down to the north. Between the A533 road [682 673] and the King Street Fault near Croxton Hall, the throw is probably around 100 m. West of the Winsford Fault evidence from the Over Borehole and from Winsford Nos. 5 and 6 boreholes indicates a throw of about 50 m.

The boundary of the Wilkesley Halite and the course of the King

**Figure 21** Sections showing the structure of the Northwich Halite at Winsford

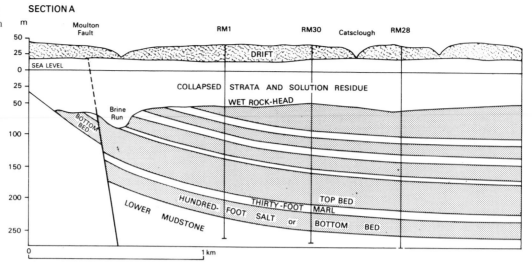

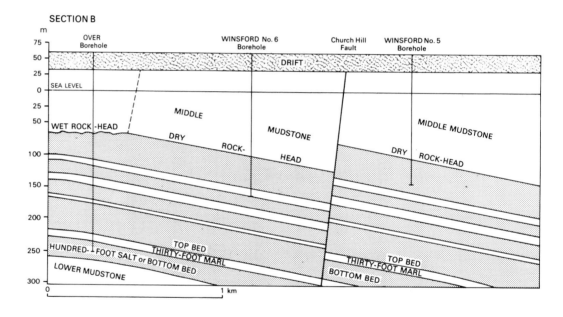

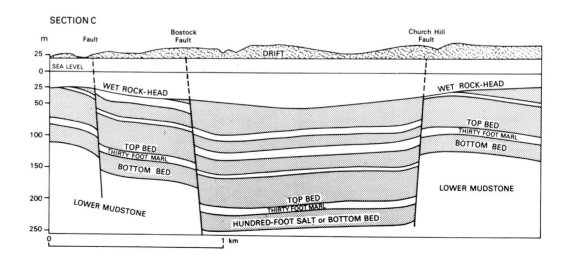

Street Fault (Figure 18) are slightly different from their postulated continuations on the current one-inch map of the Macclesfield (110) district. The new positions take into account subsequent drilling by ICI.

## FOLDS

The dominant dip of the Permo-Triassic strata of the district is gently eastwards at between 1° and 10° towards the axis of the Cheshire Basin. Local steepening of the dip is usually due to tilting between faults, or curvature of beds close to faults.

Apart from broad areas where the strata are almost horizontal, the only notable deviations from the prevalent easterly dip are in the faulted Helsby-Manley Outlier, within which the beds are synclinal along a WSW–ENE axis, and in the Peckforton Hills area where there is a very gentle syncline trending SSW–NNE. The Helsby-Manley structure cannot be detected outside the faulted flanks of the outlier (see Figures 6 and 8). Its northern limb dips southwards at about 5° and the steepest part of its southern limb dips northwards at 15° to 20°. The synclinal structure that gives the Peckforton Hills their characteristic profile is clearest between Raw Head and Bulkeley Hill on the northern margin of the Nantwich (122) district (Poole and Whiteman, 1966, p.52), but the axis continues northwards, passing west of the Peckforton Gap and through Waste Hill. It cannot be traced northwards beyond the Stanner Nab Fault. Maximum dips on both limbs of the syncline are of the order of 5° to 10°.

# CHAPTER 6

# Quaternary deposits

More than nine tenths of the Chester district are underlain by Quaternary deposits, which include glacially-derived boulder clay with associated sand, gravel and laminated clay, as well as river and estuarine alluvium, and peat (Figure 22). The chief topographic effect of the drift sheet is to smooth out the pre-glacial surface by filling the hollows and thinning against rising ground. The bedrock surface beneath parts of the Dee valley is below sea level (Wills, 1912); beneath the Gowy valley near Bridge Trafford it is more than 30 m, and beneath the Weaver valley at Northwich more than 60 m below sea level. But for the drift, parts of the Gowy and Weaver valleys would be estuarial arms of the sea. For most of the drift-covered area there is no reliable information on drift thicknesses, and therefore none on the levels of the drift base. As explained on p.65, many existing borehole records cannot be relied on to give this information, and for this reason no attempt is made in this memoir to draw a comprehensive contoured map of the bedrock surface in the district.

Drift-free areas include much of the mid-Cheshire ridge from the hills of Alvanley and Manley in the north, through

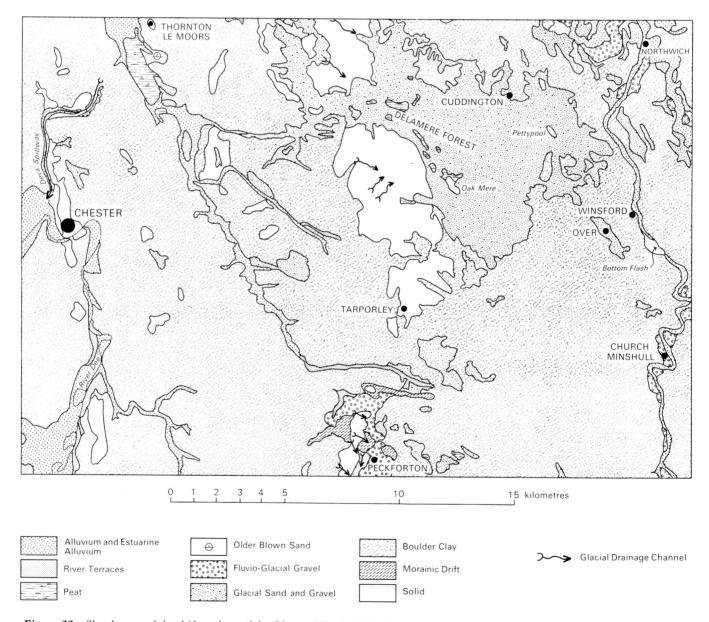

**Figure 22**   Sketch map of the drift geology of the Chester-Winsford District

Delamere and Tarporley to the Peckforton Hills in the south, the Pebble Beds outcrops of Dunham-on-the-Hill, Barrow Hill and Tarvin, and of Backford, Chester, Eccleston and Churton. There are smaller patches of drift-free Pebble Beds between the rivers Dee and Gowy, and narrow strips of Mercia Mudstone are exposed along the river Weaver and its tributaries, where valleys have been cut through the drift sequence.

Thin patches of drift, and isolated boulders, are evidence that the upland areas were once overridden by ice, but more telling is evidence outside the district, on the flanks of the Welsh and Pennine hills, that ice in the intervening lowlands must have been upwards of 400 m thick, overtopping even the heights of Peckforton. On the evidence of erratics, ice originating in Wales penetrated only the south-western extremity of the Nantwich (122) district to the south; the Chester district was affected only by ice from the direction of the Irish Sea, the junction between the two ice sheets being well to the west.

Abundant far-travelled erratic boulders of hard rock in the till are sufficient evidence of the sources of the ice-sheet in the mountains of Scotland and Cumbria, while marine shells and coal fragments in boulder clay and outwash sands record the passage of the glacier over the Irish Sea bed and the coalfields of Lancashire. Basal till is rich in locally-derived debris, mainly red Triassic mudstone in the east of the district and sand from the Permo-Triassic sandstones in the west.

Research into the drifts is summarised below: it has often been controversial, and at the time of the survey there were many unresolved problems. Chief among these was whether a widely occurring tripartite sequence of upper and lower tills separated by sands and gravels implied one glacial advance or two, or indeed whether the upper till implied a re-advance. Since then there has been much research on the deposits of modern glaciers. The work of G. S. Boulton has been of particular value in this account, since he had experience of mapping drift in Cheshire before studying the deposits of modern glaciers in Spitzbergen. It is clear from his work that, apart from the Chelford and Congleton sands which are in a class by themselves, the visible drifts of at least the northern part of the Cheshire basin can be explained quite satisfactorily as the product of one glacial advance and retreat, and that is how this account deals with them.

## HISTORY OF RESEARCH

The history of research into the Quaternary deposits of the Cheshire Basin and adjoining areas has been recounted so many times that it is not necessry to give more than a brief summary here. Early descriptions include those of Trimmer (1851) and Binney (1862), but the first attempt to classify the deposits was by Hull (1864) who found a triple sequence of drift in the Oldham area and named the components Upper Boulder Clay or Till, Middle Sands and Lower Boulder Clay. This type of drift succession was found to be widespread, and Hull's sequence was used as the main framework for many accounts of the glacigenic deposits in the region, including several Geological Survey memoirs.

The widespread occurrence of marine shells in the drifts of Cheshire and adjacent areas was probably responsible for the fact that this area was one of the last strongholds of the diluvialists who averred that the shelly drifts were produced during a great submergence of the land, the unstructured boulder clays being the debris deposited from floating ice

masses. Early collections of marine shells from the drifts of the district were those of Trimmer (1833) and Egerton (1835, 1836). Very full lists of foraminifera, ostracods, bivalves, gastropods, polyzoa, cirripedes, annelids, echinoids and sponges, compiled from the papers of Shone (1874, 1878a, 1878b) are given in Strahan (1882, pp. 26–29). The realisation that these were derived forms removed much of the incentive for studying them as assemblages, although it was clear that such studies might give an indication of the age of the drifts that incorporated them.

The mechanism by which a shelly fauna, together with its mud matrix, can be lifted from the sea bed and transported for long distances in glacier ice with little damage to many of the contained fossils has, not surprisingly, been the subject of little direct observation. Debenham (1920) described how floating shelf-ice may lift blocks of undeformed marine mud by basal freezing. Thompson and Worsley (1966, p.204) considered that, once entrained, these blocks became incorporated in an ice-sheet, where they were protected from distortion unless they were in zones of shearing. Boulton (1970a, p.224; 1970b, p.242) observed unbroken fragile marine shells in the debris layers of modern Svalbard glaciers, and considered that they were first incorporated in the ice by some form of basal freezing, and then elevated in the ice mass in a series of lenses between shear planes.

Complete acceptance of the glacial origin of the boulder clays and associated sands and gravels came gradually but, by the time the six-inches to one-mile resurvey of the Cheshire Basin was started in the 1930's few opposed the general view that the drift was the result of the invasion of the area by an ice-sheet. For some, Hull's tripartite sequence of drift had time-stratigraphic significance, with the implication that it represented more than one episode of glacial activity, and this idea was given a new impetus when the authors of the Wigan (84) memoir (Jones and others, 1938) suggested that the Upper Boulder Clay of that area was the product of a re-advance of the Irish Sea ice-sheet, and that the Middle Sands were laid down in ice-dammed water during a period of retreat separating Lower and Upper Boulder Clay times. This explanation for the drift sequence was subsequently adopted in the Stockport (98) and Nantwich (122) memoirs (Taylor and others, 1963; Poole and Whiteman, 1966) and by Simpson (1959). Attempts to fit the postulated separate Lower and Upper Boulder Clay glaciations into the accepted European Pleistocene stages, or even to correlate them with the glacial episodes of other parts of Britain, gave rise to widely conflicting views that are summarised by Yates and Moseley (1967, p.117) who, with Boulton and Worsley (1965, p.704), cast doubt on the time-stratigraphical significance of the triple sequence by showing that it was in places an oversimplification of the actual succession. Coope and others, (1971, p.95) submitted evidence from a study of fossil coleoptera that there was only one major British glaciation in the Devensian (Weichselian), and that lowland Britain was ice-free for most of that Stage; only during the latest cold phase, between 25 000 and 15 000 years ago, was precipitation adequate to sustain major ice-sheets.

The key to the interpretation of the drifts of this district is thought to lie in the work of Boulton (1972) who has shown, on the basis of surveys of the deposits of modern glaciers in Spitzbergen, that till sequences such as those in Cheshire, previously explained as the result of repeated glacial advances, are closely matched by the normal products of a single retreat phase. Serial sections through the drift deposits in the north-eastern part of the district (figures 23 and 24) show a striking similarity to the

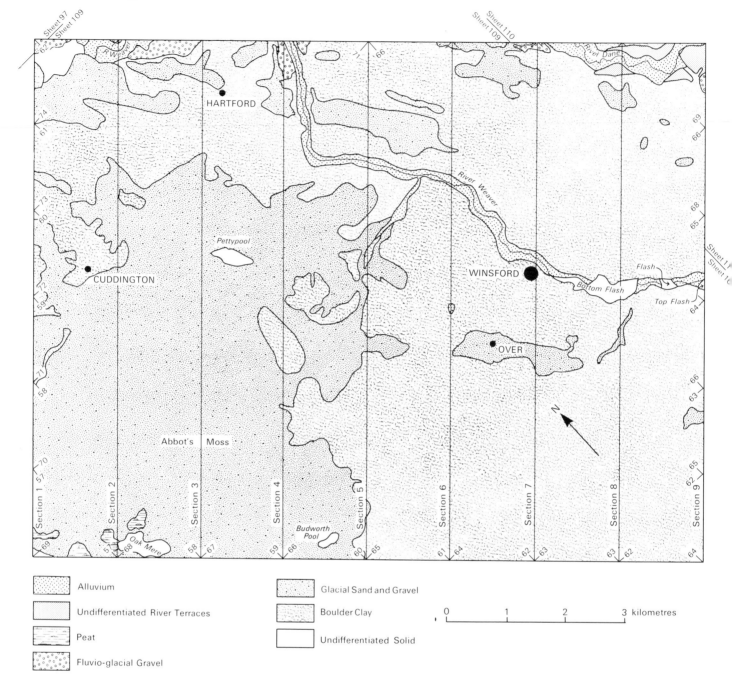

**Figure 23**   Drift deposits of the Winsford area

Alluvium

Undifferentiated River Terraces

Peat

Fluvio-glacial Gravel

Glacial Sand and Gravel

Boulder Clay

Undifferentiated Solid

0   1   2   3 kilometres

composite sections of modern glacial deposits shown by Boulton (1972, fig. 3). Two important aspects of the production of glaci-genic sequences described by Boulton are the role of ice-cored moraines, and the behaviour of flow tills. Their importance has been insufficiently stressed in the interpretation of British drifts. In this account it is accepted that most of the drift of the district, and all that can be mapped at the surface, can be explained in terms of one Devensian glacial advance and retreat.

## CHRONOLOGY AND CORRELATION

An early contributor to the debate on the age of the drifts of the

region was Reade (1888), who attempted to estimate post-gla-cial time by the amount of erosion in the Mersey, Dee and Rib-ble valleys. Following W. B. Wright's (1914) differentiation of 'older' and 'newer' drifts in Britain, it was generally agreed that the glacial deposits of Cheshire belonged to the 'newer' phase because of their relatively fresh topographic appearance.

L. F. Penny (1964, p. 407) expressed the view that the Newer Drift maximum of Great Britain might well belong to the Main Würm, and not the Early Würm as had been generally sug-gested, and Evans and others, (1968, p. 253), out of several in-terpretations of the chronological evidence, also favoured that view. Some of the radiocarbon ages recently obtained on material from sites in or near this district are given below. The overall result of these findings is to support Penny's hypothesis.

At the BGS Burland Borehole [6018 5333], about 2 km south of the district boundary, an incomplete core of grey silty clay between 27.4 and 37.5 m below surface yielded plant remains and beetle fragments (Poole and Whiteman, 1966, p. 106). The bed immediately underlay the local drift sequence of boulder clay and sand, and rested on 0.6 m of reddish brown stony clay overlying Triassic mudstones which bore signs of salt solution-subsidence. The beetles were determined in the Geology Department, University of Birmingham, and indicated a temperate climate. The plant material proved to be radioactively inert, and was given a minimum age (Birm. 57) of >38 000 years (Shotton and others, 1969); it was believed to be much older than the Upton Warren (Gottweig) Interstadial in the middle of the last glaciation, and it was thought that it could even be Ipswichian (Eemian). This implies a Devensian (Weichselian) age for the glacigenic sequence overlying the organic clay, a sequence continuous with that in the Chester district.

About 13 km east of the district and 1 km south-west of Chelford, a layer of organic mud in the Chelford Sand (Evans and others, 1968, p. 211) was found by Simpson and West (1958, p. 239) to contain an assemblage of plant remains, including driftwood and pollen, indicating a cool period characterised by the growth of coniferous forest. Radiocarbon dating on driftwood by De Vries of Groningen University gave an age of 57 000 years, later amended (Coope, 1959, p. 89) to 60 800 years. From the same deposit Coope listed a beetle fauna indicating a boreal forest environment with a temperate climate, and resembling the modern fauna of south central Finland. The three lines of evidence are consistent with deposition in an interstadial in the early Devensian (Weichselian), before the Upton Warren (Gottweig) Interstadial. The Chelford Sand, which was interpreted by Evans and others (1968, p. 248) as a fan of largely wind-blown sand derived from the east, is unconformably overlain by a sand and till sequence which equates with the drift mapped in the Chester district.

A radiocarbon assay made on marine molluscan shells collected from coarse sands at the base of a till/sand complex ('Upper Boulder Clay') at Sandiway gave an age of 28 000 years + 1800, - 1500 (Boulton and Worsley, 1965, p. 706). The shells, glacially derived from the Irish Sea, are clearly older than both upper till and the underlying Delamere Sand. These deposits are the product of the last glacial retreat in the area, and the dating reinforces their correlation with the maximum of the Weichselian glaciation (Main Würm Maximum) which followed the Paudorf interstadial in continental Europe. A younger age limit of 10 000 years BP is given to a till in the Delamere Sand at Hatchmere by the discovery of Zone II organic deposits overlying it.

Finally, B. M. Rowlands (1971, p. 10) reports an age dating made on a mammoth carpal bone originally collected from a cave in the Vale of Clwyd in north Wales by H. Hicks (1886). The Clwyd caves were said by the excavators to have been sealed by undisturbed 'northern' boulder clay, a probable correlative of the Cheshire drifts. The radiocarbon date (Birmingham 146) of 18 000 years + 1400, - 1200 is regarded as reliable, and falls within the range 25 000 to 14 000 years BP that is widely held to be that of the late Devensian, in Britain.

To sum up, the accumulated evidence strongly supports the view that the glacigenic sequence of this district is of late Devensian age — the time-equivalent of the Main Würm Maximum of the European continent. In the view of Coope and others (1971, p.95), this was the only major British glaciation in the Devensian Stage (see p. 63).

## PRE-DEVENSIAN GLACIATIONS

Deposits from early or pre-Devensian glaciations are most likely to exist, it they survive at all, in drift sequences occupying topographical depressions in the pre-glacial rock surface, like the sequence at Burland described above, although there are no such dated borehole records from this district. Records of thick drift sequences which might be thought likely to contain early-or pre-Devensian drifts include those pierced in boreholes at Austin's Springs, Newbridge (No. 4), Ways Green and Stocks Stairs; all these records, however, were made when drilling techniques in unconsolidated sediments were inadequate to provide reliable samples from accurately known depths, and the samples obtained were not always reliably described. For instance, Sherlock (1921, p. 32) used the drift record of Newbridge No. 4 Shaft [6577 6881] as evidence of a pre-glacial channel, whereas it now seems likely that the bottom 44.8 m of this bore was not 'boulder clay and gravel' but collapsed Triassic mudstone and solution residue above a wet salt rockhead (see p.43). This reassessment casts doubt upon the reliability of other old provings of thick drift including the Austin's Springs Borehole [5952 6722], which was also used by Sherlock (1921, p. 21) as evidence of a buried channel, and the Ways Green and Stocks Stairs boreholes.

Great caution is also essential in interpreting geophysical logs of boreholes through drift overlying collapsed and brecciated strata in the salt sub-crop area. Voids are introduced into the sediment pile during collapse, and are not entirely removed by subsequent re-packing. The result is to reduce the effective bulk density, to introduce water into strata that were previously dry, and to affect the bulk compressive strength of the deposit. Apart from changing the engineering properties of the rock (see p.87), the physical characteristics that are detected in borehole logging are affected also, moving them generally towards the characteristics of drift; thus, the drift base, estimated from geophysical logs alone, may be placed too low, and hollows in the bedrock surface may be wrongly diagnosed. It is partly for this reason that no attempt is made in this work to draw a comprehensive contour map of the bedrock surface.

In the Macclesfield (110) district to the east, Evans and others (1968, p.246) found residual patches of boulder clay that underlie the Chelford Sand and are therefore early Devensian or older, and the older deposits at Burland have already been discussed. Firm evidence of any glaciation earlier than the late Devensian has, however, yet to be found in the present district, and it must be stressed that multiple till layers in drift sequences do not in themselves constitute such evidence.

## INTERPRETATION OF QUATERNARY DEPOSITS

Taking into account the similarity of the Cheshire drifts to their modern analogues, and the relevant chronological facts, it is suggested that the events which produced the Quaternary deposits of the district were as follows:

A major glacier, incorporating ice from Scotland and the English Lake District, advanced by way of the Irish Sea basin and Lancashire lowlands into Cheshire, reaching a maximum about 20 000 to 18 000 years ago. It covered the whole of the Chester district and advanced at least as far south as the Bar Hill – Whitchurch-Wrexham morainic complex some 15 km south of the district (Poole and Whiteman,

1966, p.68 and plate 8). The distant origins of the ice are revealed by its content of durable boulders, chiefly of granite and volcanic tuff, and recognisable material from intermediate distances includes marine shells and shell fragments from the bed of the Irish Sea and comminuted coal from outcrops in the Lancashire coalfield. The ground-moraine or basal till reflects to a great extent the material over which the ice had most recently passed, and is sandy to the west of the mid-Cheshire ridge where it overlies the Sherwood Sandstone Group, and clayey to the east with much included Mercia Mudstone debris.

Deposits from earlier in the Pleistocene may have been preserved here and there in topographic depressions during the advance of the ice, but it appears that most pre-existing drift deposits were either eroded away before the last glacial advance or were incorporated into the deposits of the glacier. This ice combined with that emanating from the Welsh mountains to fill the whole basin from Wales to the Pennines to a depth greater than 400 m.

When the waning phase of the glaciation began the ice-front receded relatively rapidly, and thick masses of motionless (stagnant) ice covering the low ground were left to decay, detached from the parent glacier to the north. These stagnant ice-masses became thickly covered by till, partly derived from the original supraglacial till and partly from englacial debris released from the down-wasting ice, forming ice-cored moraines with considerable topographic relief. Major englacial and subglacial drainage channels that had developed during the active period of the ice were preserved as narrow channels through the ice-cored moraines. Meltwater flowed southwards in these channels towards the Severn catchment, possibly by way of Adderley, Market Drayton and the Ironbridge Gorge, being ponded on the way into temporary lakes (see Poole and Whiteman, 1966, pp.73–77).

Minor halt stages in the recession of the active ice-front are marked by terminal moraines in the Burdwardsley-Peckforton-Beeston area, but the main pause in the recession to leave its mark in the deposits of this district was at a line from Kingsley through Norley to Cuddington. When the ice-front to the east of the mid-Cheshire ridge was at this position the ice on the west side had not receded so far, possibly due to the influence of Welsh ice to the west, and the edge of the active glacier lay along the west side of the ridge, close to the position shown in the Nantwich memoir (Poole and Whiteman, 1966, figure 12) as the '260-250 foot Lake Lapworth Stage'.

Drainage waters from the ice-front along the Kingsley–Cuddington line was joined by that from the western ice-lobe flowing eastwards through a wide gap in the mid-Cheshire ridge at Mouldsworth. The combined waters were impounded in a roughly triangular area between the ice-front to the north, the northern edge of the ice-cored moraine between Little Budworth and Moulton to the south, and the combined barrier of the rock ridge and the ice-front to the west. The main outlet southwards through ice-cored moraine was by way of Over, Church Minshull and Crewe in a channel that probably originated as a subglacial or englacial tunnel. A second channel apparently operated at the same time from Oulton Park by Towns Green and Poole to the southern edge of the district.

The vigorous glacial streams cut through the basal till to bedrock in places. Later the channels were filled up with sandy outwash material. The drainage regime lasted long enough for the Delamere triangle, and the overflow channels, to be filled with sediments, mainly sand and gravel, up to levels ultimately controlled by outlets far to the south: these sands are informally termed the Delamere Sand in this account. The highest, and presumably amongst the earliest, of the outwash sands lie at 120 m above sea level near Maiden's Cross between Alvanley Cliff and Birch Hill, but the general summit levels of the Delamere Sand are about 90 m above sea level immediately east of the Mouldsworth gap. Eastwards the summit levels fall off gently to around 60 m in the vicinity of the Whitegate to Sandiway road. At Marton there is a perfect gradation at about 67 m from the generalised upper surface of the Delamere Sand into the summits along the ridge of Over Sand (see p. 74), which fall gradually along the ridge crest to 66 m at Over, 58 m at the Ash Brook crossing, and 52 m about 1.5 km south of Church Minshull, at the eastern edge of the district. There was also at about this time, or perhaps somewhat earlier, a substantial glacial outwash stream flowing eastwards from the ice-front through a gap in the ridge north of Beeston Castle now occupied by the river Gowy, and thence south-eastwards along a channel through the ice-cored moraine by way of Calvely Hall and Barbridge to the southern limit of the district at Stoke Park. The resultant outwash deposits of sand and gravel have summit levels of around 72 m at Beeston and Tilstone Fearnall, falling to about 69 m at Alpraham and progressively to around 61 m at the southern limit of the district.

So long as there was a vigorous flow of water through the proglacial drainage system any flow till sliding down into the channels from the slopes of the ice-cored moraine on either side was winnowed out with the other fines, and only the coarser material was left. When these channels ceased to operate, deposition of outwash sediments also ceased, and the semi-fluid till formed a layer over the outwash deposits in the channels. The northern and southern slopes of the Delamere Sand mass also received their cover of till at this time. When the ice inside the moraines wasted away, the channel sediments with their capping of flow till were left as upstanding ridges, while the ground between became featureless plains of melt-out till and flow till. The situation is similar to the inversion of topography in the morainic areas of modern glaciers caused by the melting of ice-cores (Boulton, 1972, p.378). Removal of support at the edges of the channels caused some slumping of the sediments, with small-scale faulting. Small faults in the Delamere Sands, and some of the self-enclosed hollows, are probably due to the melting of ice buried beneath the outwash deposits. Till layers contained in sequences of outwash deposits, like those mapped in the Delamere Sand and proved in the Southley Common Borehole, are probably due to a temporary variation in the drainage pattern which allowed flow till to remain in place long enough to be covered by further outwash material. Laminated clay, like that in the sand sequence at Dalefords, represents temporary ponding on the sediment surface.

The deglaciation of the district west of the mid-Cheshire ridge was probably by down-wasting of stagnant ice; there is no evidence of the progressive retreat of an active ice-front across the area after the halt phase along the mid-Cheshire ridge. The form of the drift deposits is similar to that on the east side of the ridge and south of the Delamere Sand, i.e., outwash sands and gravels generally overlain by or encapsulated in till. The chief difference appears to be the relative rarity of records of a basal clay till intervening between sands and bedrock. Whilst there may be some room for differences because of the varying incidence of borehole data, a reason might be that the bedrock on the east is exclusively mudstone, whereas that on the west is, except for the Carboniferous areas around Milton Green and Trefalun Meadows, exclusively Triassic sandstone.

When this district was largely occupied by stagnant ice in the form of ice-cored moraine it is reasonable to suppose that ice in areas to the south and east had already wasted away, and the surface in those areas had been lowered to present-day levels. Water flowing southwards through the channels in the remaining ice-cored moraine was ponded into lakes in thse newly-produced lowlands, laying down there the sheets of 'Upper Sand' and fluvio-glacial sand and gravel that are so conspicuously lacking in the Chester district. The sediments in the sand-cored ridges all show a gentle fall southwards, and it may turn out that their grading bears some relationship to lake levels and fluvio-glacial deposits mapped to the south: such relationships have not yet been determined.

## DETAILS

### Basal and buried till (lower boulder clay)

*West of the mid-Cheshire ridge*

North and east of Chester the presence of till beneath glacial sand and gravel is suggested by the log of a borehole at Chester Zoo, and of three others near Mickle Trafford and Guilden Sutton. At Chester Zoo [415 704] the driller's record may be interpreted as: boulder clay 7 m; on glacial sands 14.5 m; on 'marl', possibly a till 3 m; over Pebble Beds 29 m. Near Mickle Trafford a borehole at Plemstall Pumping Station [450 700] apparently pierced: boulder clay 9 m; on glacial sands 23 m; hard sandy clay, possibly a till 1.5 m; and Pebble Beds 73.5 m. The driller's record of a borehole at Cherry Tree House [449 689] is interpreted as: boulder clay 7 m; on glacial sands 5 m; clay, possibly a till, 5.5 m; and Pebble Beds 7 m. Near Guilden Sutton the record of a borehole starting from the bottom of a well 10 m deep at Byatts House Farm [451 678] is: sand 4.75 m; on clay and marl with thin layers of sand 9.75 m; marly clay, possibly a till, 9 m; and rock 12 m.

Strahan (1882, p.16) was informed that at Chester, outside Watergate and outside the city walls to the south of the gate, a sewer trench had revealed stony and rocky red clay resting on bedrock and separated by 'a line of sand or gravel' from 'an upper clay of the ordinary type'. He also mentions (p.16) that in the railway cutting at Newton [420 680] a thin seam of rocky red clay resting directly on bedrock was observed to underlie sand and gravel.                    JRE

*East of the mid-Cheshire ridge.*

In the valley of Cuddington Brook, near Brook House Farm [597 730] the thickness of the Delamere Sand tapers rapidly towards the northern edge of the sand-mass, until it amounts to no more than a thickish bed in the bankside. Here a lower till is exposed below the sand in the valley floor.

A small stream, tributary to the river Weaver, cuts through the NW-trending glacial sands ridge [629 732] 200 m west of Beach Hill and exposes a lower till, here a red boulder clay with small erratic boulders (Figure 24). The exposure continues along the south side of the Weaver valley almost to Northwich, the base of the overlying sand being marked in places by a seepage line.

Pettypool Brook cuts through an upper till and the Delamere Sand to expose a lower till from near Bark House [636 688] to the confluence with the Weaver. South-westwards from Bark House, along Bogart Brook, a tributary of Pettypool Brook, a lower till appears to crop out, but the position is confused, especially near a subsidence lake [632 682], by the fact that the till overlying the Delamere Sand has dropped to stream level through ground collapse. Figure 24 shows the conjectured section through this ground.

In the banks of the river Weaver south of Northwich, boulder clay cannot be differentiated into upper and lower layers except where glacial sands intervene. Where a lower layer can be recognised it is unmistakably a till, and contains, besides far-travelled erratics derived from the north, much material from the Mercia Mudstone, especially in its lower layers. The typical deposit can be seen in small gullies in the vicinity of the main line railway bridge over the river Weaver [6425 7066]. Here boulder clay, with well-rounded erratic boulders and pebbles, and much Triassic mudstone debris, can be seen beneath glacial sands to the north of the railway and above Mercia Mudstone to the south.

Southwards to Church Minshull, at numerous points where glacial sands provide a separating stratum, lower till can be seen cropping out low down on the valley sides. South of the railway viaduct across the Weaver, solid rocks sporadically occupy the lowest few feet of the valley slope, and a lower till can be seen resting on them. Near Magpie Lodge [639 705] in Vale Royal Park, for instance, about 50 cm of till rests on red and grey banded sandy mudstone, and this till is apparently the lateral equivalent of that underlying the western feather-edge of the sand at Moulton [6423 7044] on the eastern side of the Weaver. At least the lower part of some 15 m of boulder clay seen resting upon red mudstone in a gully [6475 6955] feeding into Vale Royal Cut is probably a lower till lying stratigraphically below the Moulton sand to the north-east, and also below the Delamere Sand, the extreme eastern edge of which is exposed at two localities [645 693; 646 690] on the west bank of the river (Figure 24.5).

In the town of Winsford solid rocks are exposed low down on both sides of the river as far south as the main A54 road-bridge over the Weaver, but exposures of the basal drift are not good. Soil indications are that the lowest drift along the valley is generally boulder clay, although glacial sand rests directly on Mercia Mudstone in a gully [656 670] running from Wharton down to the river (Figure 24.7). South of the town, exposures in gullies on both sides of the main valley in the neighbourhood of the Flashes reveal a lower till whose field relationships place it between solid rocks below and glacial sands above. The position is well illustrated in Firwood Brook [656 645]. (Figure 24.8). The sand here is the eastern feather-edge of the Over Sand, and a similar relationship can be demonstrated between basal till and the sand edge in a gully south of Church Hill Farm [649 652] and in Badger Wood [660 642].

Ash Brook cuts through the drift complex to solid rock in the vicinity of Ashbrook Bridge [656 626]. West of the bridge a basal till resting upon Mercia Mudstone is progressively cut out towards the axis of the sand lens where the Over Sand rests directly on solid rock (Figure 24.9). Farther west along the brook the lower till reappears. East of Ashbrook Bridge the basal till is red with small rounded and, in places, heavily weathered erratics of the 'Irish Sea' suite.

To the north and east of Church Minshull the river Weaver cuts through the Over Sand, to a basal till resting on solid rocks. On the outside of a bend in the river [6635 6155], boulder clay infills a trench near river level in the surface of Mercia Mudstone and is overlain higher up the bank by sand. East of the village the lower till

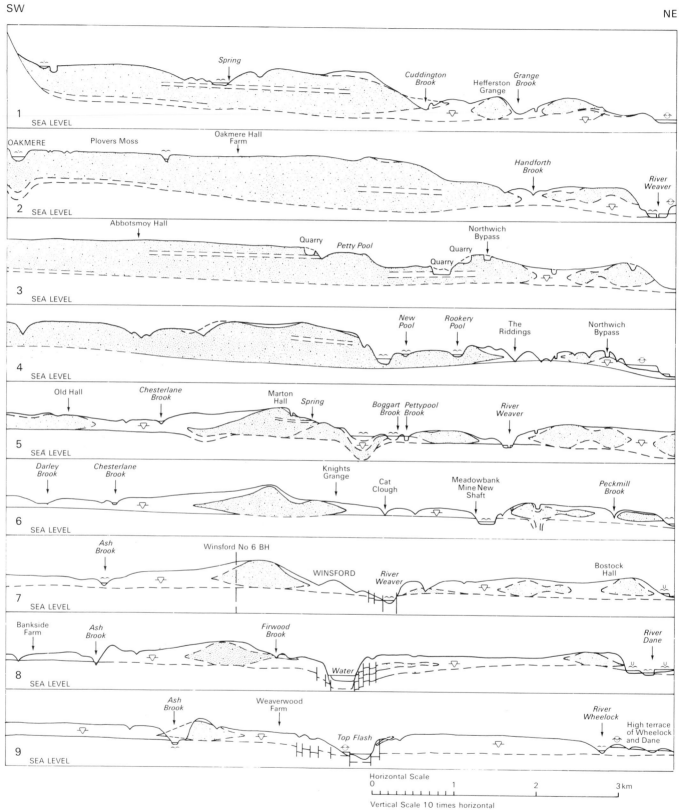

**Figure 24** Sections through the drift deposits of the Winsford area

is not exposed, but its presence between the Over Sand and the solid rocks near river level is inferred from soil indications. The mapping does not establish whether the till is continuous or partly removed along the axis of the sand feature as is the case farther north.

Several boreholes east of the mid-Cheshire ridge have proved a basal till beneath stratified deposits. The logs are summarised in Appendix 1.

In the Oakmere area, the Oakmere and Bowyers Waste boreholes proved around 5 m of stony till beneath thick sands. At Fourways Cafe [5654 6861] a record of 'hard marl and stones' between a clean gravel above and the Sherwood Sandstone below is interpreted as till. These records are the basis of the lower clay shown in Figure 24 at the south-western ends of sections 1, 2 and 3. The Over and Winsford No. 6 boreholes proved a basal till beneath thick stratified sands. The latter record is derived from the gamma log through the drift. Both sites lie on the ridge cored by the Over Sand, but a short distance south-west of the ridge axis where the Over Sand is shown as transgressing downwards through the lower till, to rest on solid rock near its axis (Figure 24.7). There is some evidence for this where the sand is deeply dissected by Ash Brook north of Church Minshull (see p.75), but it is conjectural elsewhere.

A basal till is apparently absent at the Bryn No. 2 [6022 7253] and Crabtree Green [5797 7084] boreholes, the clay between 12.19 and 17.07 m at Bryn being regarded as a flow till (see p.66). At these sites a stratified outwash sequence rests directly on solid rock. The Southley Common Borehole [5816 5934] proved a sand and clay sequence some 27 m thick resting on 8.23 m of more or less sandy red-brown till with erratics of the 'Irish Sea' suite. Much Triassic mudstone debris was incorporated into the lower part of the bed. A higher clay bed between 19.20 and 21.03 m in the middle of a sand and gravel sequence may well be a flow till (see Appendix 1, p.107).

At the Worleston Borehole [6585 5749], an upper till was separated from a basal stony clay, 1.82 m thick and thought to be a till, by 11 m of stoneless reddish brown clay. The record of the Wardle Borehole [6063 5713] shows, beneath about 9 m of plastic and almost stoneless clay, 4.6 m of sandy clay and gravel with erratic boulders; sample recovery was not good, and it is thought that part at least of the sandy clay and gravel may represent a basal till.

## Superficial till (upper boulder clay)

### West of the mid-Cheshire ridge.

Between the western edge of the district from Whitbyheath to Eccleston, and the drift-free parts of the hills from Newton in the north to Tarporley in the south, the boulder clay shown on the map overlies glacial sand and gravel or, where this is absent, rests directly on the solid rocks. Its thickness does not normally exceed about 7 m in this area, but it is recorded as about 14 m in a borehole [428 746] at the Helsby Cable Works, and as nearly 10 m in a borehole [483 743] at the former Helsby Waterworks. The relationship of the upper till to the glacial sands, where these are present, is described under the detailed account of the latter formation (p.72). Almost all surface sections within this area are of 2 m or less of reddish brown, or red and grey mottled, stony clay, which varies in texture from stiff clay to friable silty or sandy loam. The stones are mainly pebble- to cobble-sized, and include the usual great variety of erratic rock types — volcanic, intrusive, sedimentary and metamorphic. Larger boulders occur here and there in or on the clay, but the visible ones are usually by roadsides or in farmyards, having been carried there from their original sites in the fields. The largest are mainly coarse grey granite, and there are also some large boulders of Lake District volcanic rocks. Strahan (1882, p.29) lists conspicuous erratic boulders mostly located near to or on the Pebble Beds ridges of the mid-Cheshire hills, and many of these were still identifiable at the time of survey. The most arresting is a huge volcanic erratic, 3 m long, that lies [555 697] in the middle of a field

on the lower north-eastern slope of Eddisbury Hill.

At many localities over a considerable area, from the south-eastern outskirts of Chester eastwards to the Duddon-Clutton area, the boulder clay shown on the map is unusually thin, as well as being sandy in composition. It commonly includes pebbles derived from the Pebble Beds and could equally well be described as loamy sand. Such areas are fairly easily identified on a large-scale topographical map by the absence of the so-called 'marl-pits' that occur in almost every field on the more normal reddish brown clay, which was formerly extensively dug for spreading on fields.

Boulder clay (till) formerly provided a source of brick clay over a wide area on the eastern outskirts of Chester. Places where it was dug, usually to a depth of around 2 m, include Hoole Bank Brickworks [433 692] (4 m dug), small pits at Piper's Ash [434 677], north-east of Piper's Ash [438 672], by Guilden Sutton School [441 682] a former brickworks [427 672] at the end of Brook Lane, and a former brickworks [445 668] at the south side of the A51 east of Littleton. At the last-mentioned locality 7.6 m of sand was also extracted beneath about 2 m of brick clay. Nearer to the city, in areas now largely built on, Strahan (1882, p. 36) observed 'There are numerous brick-pits in the suburbs of Chester, chiefly in the neighbourhood of Boughton and Bishop's Fields. The clay is often found to be of good quality for brick-making close to its margin, where it is thinning off upon a rising slope either of rock or sand. Large quantities of bricks were supplied to the city formerly from the hollow now occupied by the cemetery, and those used in building the bridge of the Cheshire Lines across the Holyhead Railway were made on the spot. In both cases the pits were situated close to the margin of the clay, so that the rock was laid bare in the excavation'.

The railway cutting in the Chester suburb of Newton [417 678 to 425 682] is the site where Shone (1894, 1878b) first observed foraminifera and ostracods in blue silt that he extracted from the cavities of *Turritella* shells embedded in the reddish clay. Strahan (1882, p. 27) lists 61 species of foraminifera and 26 species of ostracods from this locality. As the fauna probably has no greater relevance to the glacial history of the district than that of any other fossiliferous erratic in the till, Shone's fossil lists have not been updated.                                                                JRE

South of the latitude of Eccleston, Saighton and Huxley, and west of the Peckforton Hills, the superficial till sheet is even more continuous than it is to the north. Featureless clay covers the Poulton area west of the river Dee. A well at Moat Farm [390 604] situated about 12 m above OD has proved 24 m of boulder clay resting directly upon red sandstone, showing that the drift in this area extends below mean sea level. Adjacent wells in the Flint (108) district prove that the course of the pre-glacial buried channel of the Dee hereabouts is situated within a mile of the western edge of the district. The till in this area is stiff, heavy and brown-weathering clay; the soil it makes is used as dairy pasture. Abundant Lake District and Scottish erratics are everywhere present in the clay.

South and east of Poulton the alluvium and terraces of the river Dee cover the till, which thins against the prominent north–south trending Chester-Churton ridge of Pebble Beds. South of Aldford the lowest 60 to 90 cm of till, immediately adjacent to the Pebble Beds, is very sandy and contains much locally-derived Permo-Triassic mudstone and pebbles, but the till abruptly becomes very stiff and heavy with a higher proportion of far-travelled erratics about this level. In Sourbutts Covert [412 574] the till overlies about 7.6 m of loose red-brown sand and brown silty sand with stones, possibly a sand pocket within the till. The clay hereabouts is very sandy in its lower part due to admixture with the underlying sand.

Between the Aldford-Churton ridge of Pebble Beds and the Peckforton and Burwardsley hills, an almost continuous till sheet covers most of the solid rocks, broken only by the subdued Pebble Beds hills south of Tattenhall. The clay is heavy and stiff, brown-weathering, and forming a thin soil almost entirely used for dairy farming. According to some well records in this area (Wray and Earp 1944, pp.28–29) the till in places rests on sands and gravels, which vary in thickness from 6 to 18 m. At the Tattenhall Road Brick Works pit [495 607] 6 m of stiff brown boulder clay rest on 1 m of brown stoneless laminated clay (see p.78). The boulder clay contains boulders and fragments of Eskdale Granite, Ennerdale Granophyre, Borrowdale Volcanics, conglomerates, sandstones, quartzites, vein quartz and Carboniferous Limestone.

Some 500 m SSE of Huxley Lane Farm, up to 4.6 m of stiff brown-weathering boulder clay are recorded in a small gully [521 603] draining southwards to the river Gowy, and 6 m of a similar deposit form the banks [537 604] of a small stream 410 m south of Hulgrave Farm. In the old pumping station borehole at Horton's Mill [532 602] 16 m of clay were recorded, but it seems likely by analogy with logs and exposures in the Tattenhall Road area that some of this thickness is laminated clay (see p. 78). Between Tattenhall, Handley and Chowley the drift is much thinner and Pebble Beds protrude through the till sheet as small resistant knolls, like large-scale roches moutonnées. The trend of these features is north–south, possibly reflecting the direction of bottom-ice movement during glaciation. In the immediate vicinity of the rock outcrops the till is sandy and pebbly, similar to that overlying and flanking the Aldford–Churton ridge of Pebble Beds. A short distance away from the outcrops the till rapidly becomes heavy and stiff, and Scottish and Lake District erratics predominate. Immediately east of Windmill Farm, Burwardsley [589 573], a small mound of glacial sand protrudes through the till sheet, and a larger mound forms the western slopes of the valley about 300 m south-west of Windmill Farm. The till close to these mounds contains much sand.

That the Peckforton Hills were over-ridden by ice moving from the north-west is shown by the alignment of drumlin-like forms in the till to the north-west and south-east of them. Additional evidence is the almost universal presence of a thin sandy ground moraine on even the highest parts of the hills. In general, stiff brown clay till with abundant far-travelled erratics may be traced on either side of the hills to over 150 m OD above which the thin sandy ground moraine predominates.                                   EGP

### East of the mid-Cheshire ridge

Till forms the surface of the low ground along the northern edge of the district between Kingsley and Weaverham, north of the spread of Delamere Sand. Southwards the till rises up the steep northern edge of the Delamere Sand and extends for up to 1.5 km over the sand plateau (see Figures 22 and 23). This capping till has undergone considerable dissection, with the formation of steep-sided gullies leading northwards off the sand plateau. Before this erosion the till probably formed a fairly continous sheet between Norley and Sandiway. The southern edge of this sheet is thought to represent the limit of flow tills from the ice front to the north; south of this line there is a wide area of Delamere Sand in which a till cannot be mapped, and the inference is that none was deposited.

The contact between the upper till and the sands below was seen to be irregular at a quarry [591 716] near Watermill Farm, Cuddington. Here the unstratified sandy and gravelly till was thin on the eastern side of the pit, and thickened rapidly to 3.7 m westwards, the beds below being disturbed by minor faults and thrusts. Apart from this exposure, there were at the time of survey few good sections of till overlying glacial sand and gravel in the north-eastern part of the district; the presence of the till is inferred from the nature of the soil. In the Norley and Cuddington area the fact that wherever the till overlies thick glacial sands quite minor excavations cut through it into the sands below shows shows that its thickness cannot be much more than 3.7 m (see Figure 24.1). Only the uppermost metre or so of the sides of a railway cutting [601 716] at Cuddington Station, the A49 road-cutting [602 721] by Bryn Farm, and the minor road-cutting [604 723] along Sandy Lane appear to be in clay. North-west of Bryn an old clay pit [600 730] exposed 2.4 m of stony clay on sand. The extreme eastern edge of the Delamere Sand from Hodge Lane [647 726] to Vale Royal Park is covered by a continuous strip of till which thickens rapidly downslope, and in which many clay pits, including the large Brickhill Pits [629 709], have been dug. These relationships are shown in Figure 24.2 to 5.

The south-eastern slope of the Delamere Sand is overlapped by a continuous strip of upper till as a mirror image of the till–sand relationships along the northern edge of the sand. Between Little Budworth and Marton Green [642 677] the till rises north-westwards from a featureless clay plain and becomes thinner, dying out along a highly irregular line on the summit of the sand. Detached thin patches of upper till survive around Hollybank Farm [617 687], Ballsgate [626 689], Beautybank [625 681] and west of Pettypool Brook near Bark House [636 688]. From Bark House to Parkside Farm [642 689] and thence south-westwards to Bradfordwood [637 676] the upper till, much dissected by Bogart and Pettypool brooks, overlaps the eastern extremity of the Delamere Sand.

The esker-like ridges of glacial sand and gravel in the eastern part of the district, described on pp. 74–75 and shown in Figures 23 and 24, are almost completely covered by till. Along the ridge crests the till is thin or absent, the sand showing through here and there in irregular patches. The relationship of upper till to the underlying sand was displayed at the time of the survey in a large sand pit [636 659] on the north side of Woodford Lane, 400 m south-east of St. John's Church, Over. Here 2 m of red clay till on the south-west side of the pit thinned progressively north-eastwards, towards the summit of the ridge, where it was represented by about 30 cm of a sandy unstratified deposit that would be indistinguishable from sand and gravel if the only evidence was that of the soil. The increasing overburden was the crucial factor in determining the western boundary of the pit.

Although the flanks of the sand ridges are completely covered by upper till, the edge of the buried sand lens is in places indicated by a slight change of slope at the surface. This feature can be mapped at intervals on both sides of the Over ridge, and on both sides of the Alpraham-Acton and Towns Green-Poole ridges (pp.76–78). In these examples the feature separates the low ground where the till is not know to contain sand lenses, from the rising ground where sand is present beneath a surface layer of clay. The same is true of the northern edge of the Delamere Sand, where a marked, though somewhat irregular, feature separates the low ground of Crowton, Crowton March and Onston from the rising ground to the south, and marks the northern limit of the buried sand mass.

From the low ground drained by Darley and Chesterlane brooks, SSE through Wettenhall, Wades Green and Worleston to the south-eastern extremity of the district, and lying between the Town's Green-Poole sand-cored ridges and the Over Sand ridge (see Figure 24), there is a featureless plain of clay till. This terrain is unrelieved by any of the features associated with the occurrence of major lenses or belts of sand and gravel. If there is a variable drift sequence underlying the featureless till it is impossible to infer its presence. Wettenhall Brook below Wettenhall, and Ash Brook below Darnhall Hall have cut through the till to the solid rocks below, and show it to be about 10 to 15 m thick. The clay visible at the surface is red and purple, unstratified and extremely stiff in texture. Very little of the area has been found to be suitable for arable farming, but the clay makes rich grassland. Boulders are rare, but their presence is indicated by the use of large erratics as corner-stones and smaller boulders for cobblestone pavements.                              BJT

In the remaining area east of the mid-Cheshire ridge, that nearest to the Beeston-Peckforton hills around Tilstone Fearnall, Bunbury, Spurstow and Haughton, the till forms low drumlin-like features, probably moulded around a more irregular sub-clay surface of thick glacial sands. Most of these drumlin-like forms are elongated along roughly NW–SE lines, though the orientation varies so far as to include some which trend due north and due east. The till in this area is very sandy in the immediate vicinity of the glacial sands because it incorporates some of this deposit. There are only two noteworthy sections of upper till. That at Beeston Castle sand pit [554 595], visible in 1954, showed 2.5 to 3.0 m of till near the top of the face (see also p. 77). The other section is in a pit [582 589] at the old Clayton Brickworks on the north side of the canal east of Bunbury Locks, where up to 4.6 m of light red-brown unstratified stiff clay with comparatively few stones and boulders are exposed, and a similar type of clay forms the surface of the country for 2 km or so to the south of the brickpit, as far as Haughton Hall [589 564]. At Haughton Hall two old boreholes recorded 11 m and 12.3 m of clay respectively. The clay thins west of Haughton Hall against a mound of glacial sands, and slight incorporation of the sand has occurred in the basal layers of the till.                                    EGP

## Morainic drift

Around the Peckforton and Beeston hills three large elongate masses and one small ridge of morainic drift have been mapped. They are believed to be end-moraines marking minor halt stages in the retreat of an ice-sheet. Their association with glacial drainage channels and stages in the development of glacial lakes to the south have been discussed in the memoir on the Nantwich (122) district (Poole and Whiteman, 1966, pp.75–77, figs. 11 and 12). For convenience they are named (i) the Burwardsley Hill Moraine, running for about half a kilometre south-eastwards from the east side of Burwardsley Hill; (ii) the Burwardsley–Peckforton Castle Moraine, a mass on the north-west side of the Peckforton Hills, from which a spur runs south-westwards for a kilometre or so to Burwardsley village smithy; (iii) the Peckforton Hall Moraine, a dissected mass on the east side of the Peckforton Hills, separated by the headwater brook of the River Gowy from a well-defined ridge around Peckforton Farm that trends from north-west to south-east; and (iv) the Beeston Castle Moraine, a long ridge running for more than 2.5 kilometres almost due east from Beeston Castle hill.

### Burwardsley Hill Moraine

This moraine extends south-eastwards from fields about 200 m south of Burwardsley Hill Farm [512 559] for a distance of half a kilometre and ends against the Helsby Sandstone scarp of Little Heath (Sheet 122). All but the north-western tip lies within the Nantwich district where its maximum width is about 140 m. It exhibits a steep ice-contact face on its north-eastern side, where it consists mainly of clay with stones, and a gently graded south-western outwash slope where the ground is sandy. The crest of the moraine is composed of sand, sandy clay, clay and gravel indiscriminately mixed.

### Burwardsley-Peckforton Castle Moraine

This moraine is in two parts; (a) a large, roughly triangular mass along the western slopes of the Peckforton Hills, extending from below Peckforton Castle to Spring House [527 571] in the south, and to the Crimes Brook boundary of Pennsylvania Wood in the west; (b) a ridge about 100 to 200 m wide, that trends in a south-westerly direction for over a kilometre from Crimes Brook and tapers away in the fields about 200 m south-west of the smithy [513 570] at Burwardsley village. The first mass consists of sandy clay with ridges and mounds of loose sand and gravel. Patches of very sandy boulder clay occur, and some 3 m are visible above sandstone in the track side below the Peckforton Castle escarpment. Some of the drift boulders hereabouts are rammed into northerly-facing bedding-planes in the Upper Mottled Sandstone, showing that the direction of bottom-ice movement was from the north along the line of the Peckforton Hills. The second mass, a well-defined ridge, consists of fairly stiff sandy brown clay with erratics and large sand pockets. Some 150 m north-west of Outlane Farm [521 573] up to 3 m of loose sand and gravel, part of a pocket, is visible by an old marl pit that exposes 3.5 m of stiff brown sandy clay with stones. About 450 m south-west of the marl pit other old excavations for sand [516 572] occur along the ridge, and in the road-cutting west of Burwardsley smithy up to 2 m of sand and sandy clay are exposed. The ridge has a steep north-west ice-contact slope, and outwash material was noticed on its south-east slope.

### Peckforton Hall Moraine

The main ridge around Peckforton Hall Farm [545 566] has a steep north-east ice-contact slope and a gentle south-west slope associated with fluvio-glacial deposits of sand, silt and gravel. It is composed mainly of sandy stony clay with scattered sand pockets. This mass was probably originally continuous with the dissected area of morainic drift on the east slopes of the Peckforton Hills north of Peckforton village, but has since been separated by the erosion of the drainage channel now occupied by the headwater brook of the river Gowy. The dissected mass consists of light sandy loam with gravel, sandy clay and stiff brown boulder clay. Up to 4.5 m of such material are visible in the old pit [536 571] 275 m WNW of Fountains Cottage. It has been greatly modified in shape, and probably re-sorted by water. The interpretation of this moraine in relation to neighbouring outwash gravels, believed to have accumulated in a lake at 93 m OD is illustrated in Poole and Whiteman (1966, figure 11).

### Beeston Castle Moraine

This consists of a low ridge 100 to 300 m across, which extends eastwards from Beeston Castle Lodge for a distance of about 2.6 km. The ridge is composed mainly of sandy clay with seams and patches of coarse gravel and silty sand. It rests upon boulder clay and glacial sand except at a small quarry [542 593] north-east of Castlegate Farm, where it rests on Tarporley Siltstones and glacial sands. This moraine may have been formed when an ice-front ran southwards and eastwards from a position around the western and northern crags of the hill now dominated by Beeston Castle. To the east of the hill, the ice apparently stood along the line of the moraine, and to the south meltwaters probably passed through the gap between Beeston and Peckforton castles.                      EGP

## Glacial sand and gravel

### West of the mid-Cheshire ridge

Glacial sand and gravel, mostly covered by a well-defined sheet of reddish brown stony clay underlies a large part of the low-lying, in many places almost flat, country between the western limit of the district, from Whitbyheath to Eccleston, and the rising ground of Helsby, Manley, Ashton Hayes, Kelsall and Tarporley. In the few places in this tract, where bedrock rises near to or over 30 m OD it is usually drift-free, and till laps around and rests directly on it, but elsewhere with few exceptions glacial sands intervene between till and bedrock. The only places where a surface of sandstone below 10 m OD is known to be directly overlain by till are at Hapsford, Thornton-le-Moors, along the west side of the Pebble Beds ridge of Dunham-on-the-Hill and Great Barrow, and along the lower parts of the Backford-Chester-Eccleston outcrop of Pebble Beds. In short, therefore, west of the western slopes of the Helsby-Tarporley hills glacial sands form very wide spreads which cover much of the low-lying bedrock and rarely rise above the 46 m contour; for the most part, therefore, these sands are unexposed. There is some evidence to support Strahan's contention (1882, p.16) that these spreads of sand tend to expand on the south-eastern sides of the outcrops of solid, but this relationship is somewhat tenuous.

Glacial sands are exposed in steep-sided valleys, in places on flat ground where overlying till is thin or absent, and in many sand pits. Steep-sided valleys include the valley of Backford Brook, the south side of the broad spillway between Caughall and Wervin, and the valley of Ashton Brook. Smaller ravines include 'The Dungeon' [416 725] at Croughton, and similar though less spectacular channels at Picton, Upton, Mickle Trafford and Guilden Sutton. At the latter, most of the old village is built in the channel. Glacial sand and gravel with little or no cover of till crops out at Wimbolds Trafford, Bridge Trafford, Newton-by-Chester, Hoole, Christleton, Rowton Moor, Ashton, Tarvin Sands, Stapleford, Duddon, Burton and Clotton.

A large sandpit on the west side of the A41 at Backford [398 712] exposed reddish brown grey mottled stony clay 3 m, on sandy loam 3 cm, brown laminated loamy clay 3 cm, hard cemented layer with stones 5 cm, over cross-bedded reddish brown sand 3 m. Another large pit [405 695] on the west side of Upton, probably that illustrated by Strahan (1882, figure 3), exposed reddish brown stony clay 2.5 m, on laminated buff sand with loamy layers 6 m. A deep pit [412 692] on the east side of Upton showed reddish brown stony clay with an irregular base 3 m, on cross-bedded buff sand 9 m. Along a north–south line across the middle of this pit the till thickens as if filling a former channel in the sand. Sandpits are notably ephemeral and many are quickly filled, leaving little or no trace of their former presence. At the time of survey there were sandpits of various sizes north-east and south-east of Lea Hall [392 714], at Butter Hill [411 707], Wervin [421 720], The Dale Rifle Range [399 698], the south end of Upton Golf Course [406 687], Mickle Trafford [443 696], Bridge Trafford [430 714], Guilden Sutton [448 682], Barrowmore Gorse [475 676], Peel Hall [498 698], Horton Hall [494 685], south of the A51 east of Littleton [445 668], at various sites around Christleton such as Little Heath [444 660], Birch Heath [450 658], South View [450 653] and Rowton Bridge [445 650], at Rowton Moor [449 648], Waverton [460 631], Cotton Hall [466 662], Stapleford Hall [491 645], Ford Farm Stapleford [482 645], Duddon Heath [507 652] and Duddon Hook Lane [509 647]. The pit at Guilden Sutton is probably the one illustrated by Strahan (1882, figure 7), and the pit at Ford Farm, Stapleford is probably the subject of the same author's Figure 5.

Along the western face of the hills between Alvanley and Kelsall the low-lying glacial sands of the western plain pass into higher-level glacial sands that lie along the slopes and overspread the thickly drift-filled upper part of the Ashton Brook valley. The overlying till appears to rise eastwards with the sands, but it becomes too patchy to be designated with certainty as forming a single stratum. An unusually elevated spread of glacial sand and gravel, consisting mainly of mounds, occupies the broad hollow confined by hills around Maiden's Cross [514 736] and Claim Farm. Here the sands reach a level of 120 m OD. There are many small exposures of glacial sand and gravel in the Common Side-Moor's Brook-Manley area, but few worth noting. Sand has been dug on the Ridgeway [510 747] south-east of Foxhill Pumping Station; in a gravelly mound [504 732] west of Crabtree Farm; on the south banks of Moor's Brook [500 728]; just west [503 726] of Manley Old Hall; just south of Ashton Brook Bridge [512 704]; and at Brine's Brow [524 707]. Between the western edge of Delamere Forest and Mouldsworth, Ashton Brook cuts through bold mound-and-hollow topography made by glacial sands, and these features are particularly spectacular on its south side between the mound known as Spy Hill [528 709] and Ashton Brook Bridge. In the nearby parkland around Ashton Hayes several deep steep-sided valleys have been eroded through till down into the glacial sands. Hereabouts the base of the till rises from a level of about 45 m OD in the west to about 75 m OD in the east, in a distance of about 1.5 km.

Along the western slopes of the hills from Kelsall to Tarporley there is only very local evidence of an eastward rise in the surface of the glacial sands, for till mostly laps around and rests directly on bedrock where this rises to 75 m OD or higher. In the valley between Willington Hall [533 660] and Wood Lane glacial sand and gravel is mapped at about 61 to 76 m OD and again south of Wood Lane in Oxpasture Wood. This level is 15 to 20 m higher than that of the nearest spreads of sand on the flat ground around Duddon and Clotton. Sand was dug at two places [534 653, 538 652] near Dickinson's Rough, and on the north side [540 647] of Oxpasture Wood. In much of the Stapleford-Tarporley-Willington triangle the upper till tends to be very sandy or loamy, and locally it only thinly covers the underlying glacial sands. In places near the hills glacial material makes sandy mounds and ridges, and one such area west of Tarporley, here mapped as boulder clay, was described by Strahan (1882, p. 20) as follows: 'West of Tarporley a barrier of sand extends across the mouth of a shallow valley in the Waterstones and has given rise to a mere basin now filled with peat. The sand is interbedded with clay in thin veins and bands of one or two feet in thickness. The whole is contorted, and the sand and clay jumbled up together'.                                                    JRE

South of the latitude of Eccleston, Saighton and Huxley, and west of the Peckforton Hills, glacial sands are very little in evidence. Surface exposures are confined to two small areas. About 1 km north-west of Churton village a small patch of glacial sand crops out in Sourbutts Covert [411 574]. Here, up to 7.5 m of loose red-brown sand with stones (mainly derived from Pebble Beds) overlain by till, can be seen in the dry valley. The other locality is close to the hills, where two small outcrops of glacial sand and gravel, overlain by till occur about 1 km west and north-west of Burwardsley village. The larger outcrop forms the western slopes of the valley that extends to Tattenhall. Several small exposures of loose brown sand with stones occur on these slopes, and sand and gravel are frequently thrown out by rabbits. The best exposure is in an old sandpit [505 571] at the northern end of the outcrop, where nearly 2 m of medium- to coarse-grained sand with stones are overlain by 30 cm of sandy brown till. The smaller of the two outcrops forms a low mound immediately east of Windmill Farm, and an old sandpit [509 573] about 70 m north-east of the farmhouse shows 1 m of fine-grained sand with stones. The overlying till is very sandy in the vicinity of these outcrops, and it is probable that they represent small sand masses within the till.                                     EGP

Glacial sand and gravel has been pierced in many boreholes for water. The maximum recorded thickness is at Mouldsworth Pumping Station where the log records 70.7 m, beneath 5 m of till. Other provings of thick glacial sands are of 34 m at the Shrewsbury Arms [449 709], Bridge Trafford, and 32 m at the British Insulated Cable Works [482 746], Helsby. More recent boreholes have proved considerable thicknesses in the low ground between Dunham-on-the-Hill and Manley, and in the Ashton area. At Manley Quarry Pumping Station [490 719], No. 1 Borehole recorded 32 m of sand and gravel beneath 8.5 m of boulder clay, while the log of the adjacent No. 2 Borehole suggests a total drift thickness of the order of 45 m. South of Ashton village the borehole at Ashton Pumping Station [505 688] proved about 41 m of drift, the bulk of which is glacial sand and gravel, and the average thickness of similar material in three nearby observation boreholes is 45 m. In the low-lying, thickly drift-covered area around Oscroft, Duddon, Clotton and Iddinshall, where the average thickness of drift seems to be of the order of 15 to 25 m, the thickest record of glacial sand and gravel is about 36 m in a borehole [5199 6430] at Clotton Hoofield.           JRE

*East of the mid-Cheshire ridge (Kingsley, Sandiway and Little Budworth)*

On either side of the deeply drift-filled gap occupied by the head streams of Ashton Brook, glacial sand and gravel deposits lap around the solid rocks of the mid-Cheshire hills near the level of the 91 m contour. The deposits are mainly disposed in mounds and ridges with intervening hollows, many self-enclosed, and these features are well developed between the western boundary of Delamere Forest and Hatchmere. This deposit is only the western-most lobe of a much larger triangular sheet of sand and gravel, the Delamere Sand, underlying the ground between Kingsley, Sandiway and Little Budworth. Between Kingsley and Cuddington the north-facing edge of the sand forms an abrupt feature, rising nearly 60 m above the plain to the north, which is floored by boulder clay. The till of the plain rides up this slope on to the top of the sand mass where it is preserved in irregular thin patches affected by late-glacial and post-glacial erosion. At the south-eastern margin of the Delamere Sand, between Little Budworth and Vale Royal there is a similar relationship between sand and upper till to that along the north-facing edge. Serial sections through the Delamere Sand in Figure 24 illustrate the situation.

The northern face of the Delamere Sand plateau is dissected into a series of bold bluffs by gullies which include those occupied by Small Brook, Cuddington Brook, and other north-draining valleys near the north face of the plateau. This dissection clearly shows up the relationship of till to sand, particularly near Harthill Bank on Gallowsclough Lane [568 717] where three small outliers of boulder clay, capping sand mounds, have been isolated from the main spread of upper till at Norley by the valley of Small Brook. East of Cuddington the northern slope is more subdued and the boulder clay cover more continuous; exposures are intermittent, occurring only where the sands show through windows in the overlying till.

Much of the Delamere Sand has a plateau-like form, notably within 1.5 km or so of the northern margin in the parishes of Norley and Cuddington. It is in this region, that the till cover is most extensively preserved. Elsewhere, particularly along the west side of the deposit close to the solid rock mid-Cheshire ridge, the surface is hummocky with numerous enclosed hollows or kettle holes. Apart from erosion effects, some of this relief may be due to original irregularities in deposition; certain depressions such as Oakmere [575 676], Marton Hole [618 674] and water-filled hollows along Bogart Brook [631 681 to 634 684] are known to be affected by subsidence (see p. 50); other lakes like Budworth Pool [598 656], Petty Pool [618 700] and New Pool [627 696] are artificially dammed. There are many enclosed hollows, however, mostly containing meres or peat mosses, which are not due to the above causes, nor are they related to any normal process of sub-aerial erosion. They include Flaxmere [556 723], Hatchmere [553 722], Blakemere Moss [551 711], Linmer Moss [547 607] and numerous others. The largest are concentrated close to the east side of the mid-Cheshire ridge. It is thought that they mark the sites of buried ice or ice-cored moraine over which the stratified outwash sediments collapsed during the final melt.

The best exposures of the Delamere Sand are in the larger sand pits. Sand and gravel were worked on a large scale by the Marley Tile Company at Long Ridge, south-east of Delamere Station. A section of the south end [564 694] of the ridge before it was entirely removed showed more than 6 m of bedded buff sand with loamy layers and lenses of coarse gravel up to 1.25 m thick and 9 m long. The dip of the bedding was away from the axis of the ridge.

East of Sandiway the north-south dimension of the Delamere Sand abruptly diminishes, the deposit itself becomes thinner, and its edge has been mapped in the northern environs of Vale Royal Park in stream gullies leading eastwards to the river Weaver [634 707, 636 702]. Here the sand is seen to be both underlain and overlain by till, and its edge can be placed within a few yards.

A sand quarry on the north side of Foxey Hill, Cuddington [584 719], showed 0.6 m of solifluxion gravel on a metre or so of gravelly sand. An irregular clay band up to 60 cm thick intervened between these beds and the main mass of sand below, 7.6 m of which was seen in the quarry. Another large sand pit [591 716] about 800 m

west of Cuddington and on the north side of Cuddington Brook, displayed some 9 m of cross-bedded sand. In the upper third of the sands were noted pockets of gravel, lenticular in cross-section and elongated in form, with flattish tops and concave lower surfaces, apparently laid down in channels during the deposition of the sands. The pebbles, all well rounded and up to 8 cm diameter, are northern erratics of the 'Irish Sea' suite. The gravels contain complete though abraded marine shells, presumably carried from the Irish Sea basin and washed out from glacial till close to where found. Both sands and gravels contain numerous small calcareous shell fragments. A persistent layer of fine red loamy sand resembling moulding sand was noted about 3 m below the highest part of the sand in the vicinity of the quarry. In a part of the quarry restored to agriculture near Watermill Farm [592 715], a bed of white silica-sand about 60 cm thick dipped southwards at a low angle, overlain by normal red sand. Highly siliceous sand was also noted about 500 m south of Watermill Farm and south of the railway, [5918 7095] and could be the same bed. Overlying the sand on the west side of the quarry, 2.14 m of very fine red stoneless sand rest on 1.22 m of sandy clay with small stones, all dipping gently towards the south and truncated towards the north and east at the base of an overlying till. The dip is not considered to be original, but a result of post-depositional distortion of the sand mass. The topmost beds of the sand, where overlain by till, are much disturbed by small faults and thrusts, and there has been incorporation of small masses of sand into the upper till.

The main quarries [620 710] at Sandiway at the time of the survey lay south of the A556 and north-east of the golf course (Pettypool Park). Upwards of 10.5 m of clean-washed, cross-bedded glacial sands were exposed, with many calcareous marine shell fragments and a few complete shells, found chiefly in gravel pockets. These pits were remarkable for the occurrence of what the quarrymen termed 'clay intrusions'—sub-conical masses of red stony till up to 7.6 m high, rising through the sands from below and forming a serious impediment to working the pit. Generally oval in plan, the pillars were ranged rampart-like in a line trending NW-SE, apparently the same trend as many of the englacial and subglacial streams. There were also isolated pillar-like masses of till in other parts of the pit. The sand surrounding the pillars had clearly been deposited when the pillars were already in place. Possibly the pillars were remnants, after violent erosion by outwash streams had removed most of the till leaving the pillars to be surrounded soon afterwards by glacial sand. An isolated lump of boulder clay a metre across near one of the pillars probably fell off into the sand as this accumulated. During the working life of the pit quarrymen had to work round these obstructions, which were left as pinnacles as the sand fell away around them. The till involved is not thought to be a basal till, but an intraformational layer, possibly a flow till from the ice-front to the north, formed during the deposition of the Delamere Sand (see Figure 24.3).

At the time of the survey large sand pits were being opened up north of the A556 near Forest Hill [619 714]. One of these was later described by Thompson and Worsley (1966). Their paper also provides a view on how delicate marine shells can be lifted from the sea-bed by a glacier, transported and then deposited relatively undamaged; it also discusses the significance of radiocarbon datings made on two specimens of the marine gastropod *Nucella lapillus* Linne, collected from the quarry (see p. 65).

Up to 9 m of current-bedded glacial sand with thin subordinate lenses of gravel were exposed in two large quarries at Dalefords, one on either side of the Sandiway to Winsford road. In the pit east of the road up to a metre of unstratified clayey sand formed a patchy cover to the sands [608 698]. In the western pit [605 699], 30 cm of banded clay and silt overlay the sands, representing relatively low-energy conditions of deposition in the dying phase of Delamere Sand deposition. Finely varved clay — almost certainly a bed within the Delamere Sand and not at its base — was seen in the floor of the western quarry. In both pits there was much bedding deformation, including faulting, affecting the whole height of the quarry faces, and a steep dip of 33° was noted in sands in the eastern pit. This deformation was attributed to ice loading subsequent to deposition by Taylor (1958, p.361), but that author now believes that this and most other bedding deformation in the glacial sands of the district is due to collapse after the melting of ice, either buried beneath the stratified deposits or supporting them from the side.

A sandpit [593 691] east of the A49 and north of Abbotts Moss Wood, worked by floating suction-dredger, showed an impersistent superficial layer of structureless sand up to 1.25 m thick resting on 4.6 m of cross-bedded sand to water-level, and at least a further 3 m below with the bedding distortion and microfaulting typical of glacial sands in this area. An abandoned pit [603 684] near Newchurch Common and about 1 km south-east of the last mentioned, showed about a metre of unstratified, mainly ferruginous sand with small distorted masses of red clay, on 1.52 m of current-bedded sand with distorted bedding. There was at least a further 3 m of sand below water-level, which at the time of survey was 2.75 m below ground surface. The stripped-off overburden contained numerous etched pebbles and dreikanter, and well-rounded boulders of the 'Irish Sea' suite. The relatively high water-table hereabouts may be due to a clay layer within the sand sequence, and not far below the base of these wet pits, similar to the clay at the bottom of Dalefords west pit (see above) and flooring a water course [602 697] south-west of Dalefords.

There are numerous examples of more or less persistent clay bands within the Delamere Sand. These were of real economic importance before the advent of piped water to most farms, because they hold up local perched water-tables in the sands, which give rise to springs where the clay crops out on valley sides. Along Cuddington Brook [585 711] just below Cuddington Pool, a clayey soil at the bottom of the valley and the presence of a spring line suggest there is a clay layer in the sands. The clay is exposed in a small gully [590 714] west of Watermill Farm. The Crabtree Green Borehole [5797 7084], drilled just to the west of Cuddington Pool in 1934, recorded 29 m of sand and gravel below the horizon of the clay. The relationships are thought to be as shown in Figure 24.1; the edge of the clay, dipping gently eastwards, crops out at the east end of the pool. Up the valley sides it is unconformably overlain by the upper part of the Delamere Sand. The clay is likely to be a flow till derived from an ice-front to the north during the deposition of the Delamere Sand. Its identification as a till and not as a stratified water-lain clay is supported by the presence on its surface of large hard rock erratics, including an andesite boulder with long dimension 1.52 m [5859 7100] beside Cuddington Brook east of Cuddington Pool, and a polished and striated granite boulder 1.2 m long [5926 7145] near Watermill Farm. A group of shallow wells [580 709] sunk in the valley bottom west of the pool and in an adjacent feeder [577 707] probably owe their copious water supply to the damming effect of the clay lower downstream, which maintains a high water-table west of the pool.

The outcrop of a clay band about 10 m higher in the sands than the last-mentioned one, crosses Cheese Hill Lane [580 712] about 150 m north of Cuddington Brook, giving rise to a spring [5797 7115] on the valley side (Figure 24.1). Other springs south and south-west of Gallowsclough Hill [5707 7108, 5689 7108] may also owe their existence to the same clay band, though it has not been mapped. The ponds of Gallowsclough Farm are thought to be dug to a perched water table above a clay band in the sands, and a clay layer at about the same horizon crops out on a hillside [580 719] near Royalty Covert.

There is evidence of a widespread clay band some 10 m below the general surface of the Delamere Sand in the north-western part of its outcrop, from the vicinity of Crofton Lodge [547 735] south-

eastwards to Harthill Bank [566 716]. A gully mainly in sand west of Crofton Lodge is floored by clay, and a wet line along the gully side marks the clay/sand junction. A large (1.2 m) boulder of acid volcanic tuff [5474 7379] lies close to the junction between the clay and the overlying sand near the bottom of the gully. Clay is mapped, mainly on the basis of the soil, along the north side of a gully feeding into Hatchmere [553 722] from the west. This clay probably extends eastwards beneath the adjacent Flaxmere [557 723]; this is a peat-filled relict mere where the water-table is some 10 m higher than the water-level of Hatchmere. Clay occupies the bottom of the northern valley side of a water course near Harthill Bank. An exposure [5613 7180] of red sandy clay with erratic pebbles shows that this clay is a till, analogous with that already described in the area of Cuddington Brook.

A till layer that was partly eroded and left in the form of remnant pillars surrounded by sand, in quarries [620 710] at Sandiway, has already been referred to (p. 73).

The Bryn No. 2 Borehole [6022 7253] proved nearly 5 m of 'brown clay with pebbles' in the middle of a sequence of sand and gravel. This clay is presumed to be a flow till at about the same horizon as that known elsewhere along the northern part of the Delamere Sand.

At Sherratt's Rough, 1.1 km west of Whitegate, clay floors a gully cut mainly in sand. A wet line marks the sand–clay junction and, at one point [6156 6918], 3 m of sandy till are exposed. At another clay-floored gully nearby, a spring marks the sand–clay junction but there are no other exposures. The top of the clay lies at about 84 m above sea level. The bed is shown in Figure 24.4, the line of which passes between the two localities.

The clay in the floor of the west quarry at Dalefords [605 699] is not a till but a laminated clay, and it lies at roughly the same horizon in the sands as the flow tills described above (Figure 24.3).
BJT

*Marton to Church Minshull*

A long narrow deposit of sand and gravel, is here named Over Sand after the locality that provided the most complete vertical section through it. The sand extends from the edge of the Delamere Sand near Marton Hall [624 675] south-south-eastwards to the western environs of Crewe, and thence south into the Stoke-on-Trent (123) and Nantwich (122) districts. Sections across the Over Sand (Figure 24.5 to 9) show that it is almost totally encapsulated in clay for much of its length. From Marton to Church Minshull its topographic expression is a low ridge, the crest of which is followed by the Over to Nantwich road; beyond Church Minshull the ridge is still discernable, but much more subdued, and its form much modified by tipping and major building developments west of Crewe. The Over Sand is the best developed and exposed of a number of similar features in the district (and probably in the Cheshire Plain as a whole), and is, therefore, dealt with here in some detail.

Although along most of its length the Over Sand is covered by till, there are good exposures at intervals which throw light on the form of the deposit and it relationship with formations above and below. Between Marton and Smith Bank [648 647] the ridge crest is some 15 to 18 m above the till plain on either side, and sand is exposed along the summit for most of this distance, the flanks being covered by till. Between Salterswall and St John's Church, Over, till completely covers the ridge, though it is probably thin over the crest.

The section measured in the now-abandoned Over sandpit [637 659], which was worked until increasing clay overburden westwards away from the ridge-crest made operations uneconomic, showed up to 2 m of sandy till resting on cross-bedded sand, which exhibits a reticulate minor fault-slip pattern of a type common in the glacial sands of the district and probably caused by the melting of supporting ice soon after deposition. The sand, 15 m of which were exposed in the quarry face, contained small pockets of gravel near the top. The quarry had probably been excavated almost down to the water-table towards the end of its life, its floor being subject to intermittent flooding.

The Over Borehole [6364 6587], sited in the quarry floor, proved a further 18.3 m of sand and gravel, resting on a lower till, in the following section:

|  | Thickness m |
|---|---|
| Over Sand | |
| Sand, reddish orange | 2.74 |
| Sand, as above, with reddish brown clay fragments | 0.92 |
| Sand reddish orange | 8.23 |
| Gravel with a little sand | 1.83 |
| Sand, light brown with pebbles | 4.57 |
| Basal till | |
| Clay, red, with erratic boulders and fragments of Triassic mudstone | 1.83 |

The basal till may thin out completely towards the ridge axis, so that the Over Sand rests directly on the Trias (Figure 24.3).

During the survey there were other sand pits to the south, on or near the ridge crest, the largest being that near the northern end of School Lane [647 648] which showed 6.7 m of sand. The pit bottom was quite dry, presumably because the till cover is breached immediately to the north-east by a gully leading down from the dell in which St Chad's Church stands to the Bottom Flash in the Weaver valley.

From Smithy Bank to Ash Brook the ridge is completely covered by an upper till, which is thin enough near the crest for pits for sand to have been dug through it at several points. Two of these near Court House [653 638, 654 636] showed 6 m of sand beneath 61 cm of till, and in the more northerly pit water stood at about 6 m below ground level.

Exposures in several of the gullies leading off the ridge towards the Weaver demonstrate the attenuation and disappearance of the sand bed away from the ridge crest. These gullies are, from north to south;
i.   a valley leading ENE from the Old Star Inn [647 650]. Here the sand bed, between two tills, thins and dies out just east of St Chad's Church.
ii.   Firwood Brook, in the sides of which the eastern edge of the sand lens can be mapped beneath an upper till in the western end of Shaw's Wood [657 646].
iii.   Badger Wood [660 642], a steep-sided wooded gully in which the feather-edge of the sand is mapped between two tills.

All these gullies breach the cover of clay till which impounds water within the Over Sand; the streams that occupy them are fed by springs, the extreme constancy of which, even in dry weather, is an indication of the large volume of sand, including both the Over and Delamere sands, that they drain. There is good reason to suppose that the edge of the rising ground of the ridge is the surface expression of the edge of the sand lens beneath the overlying till, for the line of this feature, where crossed by a gully, coincides in every case with the point at which the sand can be shown to die out.

That the Over Sand is roughly lenticular in cross-section is shown by exposures in the valley sides of Ash Brook, which cuts through the ridge from west to east, just west of its confluence with the river Weaver. In a west–east downstream traverse along the valley of Ash Brook, the first intimations of the presence of the sand lens are exposures of a thin sand bed in two tributary gullies west of Primrose Hill [651 629, 652 628] on the north side of Ash Brook. South-west of Ash Brook Cottage [656 624] on the south bank of

Ash Brook there is similar evidence of a thin sand bed. East of this the sand bed thickens rapidly towards the axis of the ridge, its upper surface rising and its lower surface descending until the line of greatest thickness is reached just east of Ashbrook Bridge. Although exposures are not everywhere good, and there may have been some misleading slipping on the steep slopes, it appears that the lower surface is transgressive, for at several points the sand appears either to rest upon solid rocks or to be separated from them only by a thin basal layer of till. East of the ridge axis the sand bed thins, and its base rises up the valley side, until it dies out [663 625] north-west of Lea Green Villa, where the enveloping layers of till come together.

South of Ash Brook the ridge is cut through obliquely by the Weaver, on both sides of which there are good exposures. Apart from the large Church Minshull sandpit [662 614] on the west side (Plate 3), exposures on the east or Minshull Vernon bank are better, for here the full sequence of drift is seen resting upon Triassic mudstone at river level. Here, too, the lenticular cross-section is displayed, and the deposit is seen to be narrower and thinner than it is between Over and Ash Brook, the maximum thickness being not much more than 9 m and the width less than a kilometre.

Farther south the sand lens has little topographic expression, and its presence is inferred only by exposures in sand pits dug through rather featureless till in the Bradfield Green area, and by exposures in the valley sides of Leighton Brook and Valley Brook in the western outskirts of Crewe, in the Macclesfield (110) district (Evans and others, 1968, p.226).

*Weaverham, Hartford, Bostock and Moulton*

In this area there are several narrow ridges cored by glacial sands and partly covered by till, that have a NNW trend. Hefferston Grange Hospital [605 735] stands on one, and a short distance to the east there are three marked sand ridges which coalesce northwards into a wide deposit on which most of the town of Weaverham is built. Of the two more easterly of these, one culminates southwards in Gibbet Hill [628 727] and the other extends beneath Beach Hill [631 732] to Hartford Grange [637 724]. A sandpit near the crest of the Hefferston Grange ridge just east of Grangelane Farm [606 729] exposed 3.7 m of loamy sand with few stones. Nearby, a pipeline trench, open at the time of survey, showed that till covered the sands on the flanks of the ridge and thickened away from the ridge-crest.

There are two parallel sand-cored ridges in the vicinity of Moulton, one being the broad ridge on which Moulton village stands and the other a narrower ridge about 1 km to the east trending through the Beehive crossroads [660 697]. The Moulton sand lens extends northwards towards Hartford, and it apears to be aligned with the Hartford Grange sand, though there is a discontinuity in the ridge, and a lack of other evidence for a distance of about 700 m in the vicinity of Woodlands [638 717]. The ridge is transected by the Weaver north-east of the Heyes Wood railway viaduct [6425 7064]. Here the mapping indicates that the sand has both a convex lower and a convex upper surface, for the sand bed is high up on the banks of the Weaver at the edges of the ridge, and the base is lower down the valley side in line with the highest part of the ridge. North-west of the river, exposures in the cutting for the A556 Northwich By-pass [644 714] showed more than 6 m of sand resting on till. Traced southwards from Moulton the ridge is indistinct beyond Moulton Hall [659 684]; a sand-cored ridge in which there is a sandpit 500 m south-east of Wharton Green [665 678] is on the same line, and a parallel ridge some 400 m to the north-east may mark the continuation of the narrower Moulton ridge referred to above. The twinned form of the Moulton sand persists for nearly 8 km from the south-eastern environs of Weaverham to Wharton (see Figure 24.2 to 7). An old sandpit [6687 6690] lies on the line of the Moulton ridge about 400 m NNE of Nun House, Wharton, and hereabouts the form of the ridge appears to have

**Plate 3**   Cross-bedded glacial sand (Over Sand), Church Minshull

been modified by subsidence at The Willowbeds [675 671]. This ridge cannot be traced farther south than the Middlewich road east of Winsford Station.

Bostock Green [671 693] stands on a sand-cored ridge elongated NNW-SSE (see NE end of Figure 24.6). Bostock Hall (NE end of Figure 24.7) lies a short distance south-west of the ridge axis. Northwards this sand extends beneath Davenham — east of which village it is exposed on the west side of the river Dane — thence towards Northwich. There is a parallel ridge on the east side of the Dane in this area, and it appears that the shallow valley between these ridges may have determined an early course of the Dane between Middlewich and Northwich. In the same way, the early Weaver seems to have been controlled by sand ridges between Church Minshull, where the river cuts through the ridge of Over Sand, and Valeroyal Locks, where it takes a sharp turn northeastwards and cuts directly through the Moulton ridge. It appears that several of the sand lenses referred to above coalesce in the vicinity of Northwich forming a widespread deposit, much modified by subsidence and river erosion, in which it is not possible to make out any characteristic forms.

The Bostock sand ridge can be traced for several kilometres south of Bostock Hall, and several old pits have provided exposures. At the time of survey the best of these was of 3 m of medium to coarse sand with stones at a pit [683 674] 200 m east of Oldhall Farm. South of here the ridge makes a pronounced westward swing through Stanthorne and towards Clive House [671 654] where its form is lost in ground much modified by salt-solution subsidence.

Stanthorne Hall [682 666] where there is an old sandpit, and Stanthorne Lodge [686 666] stand on the ridge. The house called Stanthorne [685 658] appears to be on a short branch off the main lens, or on an associated sand-cored mound, and here two pits exposed about 3 m of loamy stony sand. Nearby was a pile of erratic boulders of the Irish Sea suite.

### Calveley-Barbridge-Worleston

In this area there are pronounced sand-cored clay ridges with a NNW trend and the same lens-like cross sections as the Over Sand. The amplitude of the ridges is of the order of 9 to 11 m. One stems from the larger mass of glacial sand in the Tilstone Fearnall-Alpraham area north of Calveley, and can be traced by Calveley Hall, Highbank and Wardle Bank as far as the Shropshire Union Canal north of Barbridge. Here it appears to die out, but a short distance to the south, in the village of Barbridge, is the northern end of another ridge which carries on the same trend by Stoke Park and Hurleston, and extends well into the Nantwich (122) district to the south, where it provides the low eminence on which Acton stands, and is the source of water in many wells in that village.

A parallel sand-cored ridge, somewhat more sinuous and almost completely covered by upper till extends from the Old Covert [620 590] in Cholmondeston parish SSE by Greenbank and Cross Banks farms towards the district's southern boundary near Poole Green [636 559] where it widens out somewhat and extends southwards to underlie Rease Heath in the Nantwich (122) district. The sand is

**Plate 4**   Trough and ripple drift bedding in glacial sand, Beeston Castle Sandpit

exposed in a stream gully alongside Old Covert and in a nearby pit. There are old pits along the ridge north of Cross Banks [629 573], one of which [6305 5748] exposed 3 m of stony sand under a thin sandy till. South of Cross Banks the only significant exposure in the district is about 400 m NNW of Gates Farm [634 565] where 2 m of loamy sand were recorded beneath a till of variable thickness. Towns Green [607 613] stands on a sand-cored ridge elongated along the same trend as the last-mentioned, and carries the line northwards towards the southern tip of the main Delamere Sand near Rushton and Oulton Park. There is no direct evidence of the sand between Old Covert and the Towns Green ridge, but springs emerging immediately west of Fox Covert [615 600] suggest the presence of sand containing water under artesian pressure beneath the otherwise featureless surface of till.

A patch of sand immediately south-west of Lower Hall [641 564] is a rare case, east of the mid-Cheshire ridge, of glacial sand stratigraphically overlying the uppermost till.                                    BJT

*Tiverton-Beeston area*

Deposits of glacial sand and gravel, partly covered by an upper till, occur in this area in the form of large mounds of loose sand. South of Tiverton village [555 604] the river Gowy has cut into a mound of glacial sand over 1.5 km long, 0.7 km across, and elongated WNW-ESE. Exposures are good hereabouts and mapping is facilitated by numerous rabbit burrows which are confined to the sand, avoiding the stiffer overlying till. An old sand pit [554 602] on the east side of the A49 at Tiverton exposed 3 m of loose brown sand overlain by 2 m of sandy till; 10.5 m of loose brown sand with stones have been dug at another old pit [558 596] by the canal east of Beeston Stone

Lock. Several exposures of loose sand and gravel are recorded around Wild Boar Inn [559 592] on the south side of the river Gowy. The best exposure in this area is at Beeston Castle sandpit [554 595] where the following section was visible in 1954 (Plate 4).

|  | *Thickness* m |
|---|---|
| LATE-GLACIAL SAND AND GRAVEL | |
| Loose brown silty sand | 1.37 |
| Coarse gravel | 0.15 |
| | |
| TILL | |
| Clay, varying from heavy stiff brown clay to light sandy brown loam; contains many Lake District and Scottish igneous erratics | 2.5 to 3.00 |
| | |
| LAMINATED CLAY | |
| Brown stoneless plastic clay, violently contorted in places | 3.50 |
| | |
| GLACIAL SAND AND GRAVEL | |
| Sand, loose yellow-brown cross-bedded; a few thin seams of small gravel with coal streaks in the lower 1.22 m | 24.00 |
| | |
| ?TILL | |
| Clay, reported dug from drains in pit floor | |

Long, turreted marine gastropods (*?Turritella*) have been found in the sand and gravel. In places the gravel rests upon a bed of dark bituminous-looking silty sand and sandy silt, within which large blocks of Helsby Sandstone weighing up to half a tonne occur; their present positions are more than a kilometre from the nearest Helsby Sandstone outcrop. The clay in the pit floor is not exposed, but the owner stated that it was an impervious stony clay which created drainage problems in the pit.

East of Beeston Castle, at an old quarry [542 593] about 110 m NE of Castlegate Farm, 2 m of loose yellow-brown sand with a few rounded quartzite pebbles are overlain by 2 m of coarse gravel, a constituent of the moraine which extends eastwards from Beeston Castle (see p. 71).

### Tilstone Fearnall-Alpraham

A large dissected mound of glacial sand and gravel, elongated roughly NW-SE extends for about 1.5 km SE of Tilstone Fearnall crossroads [569 604]. Exposures of sand and gravel are common within the mound; in the old sand pit [577 598] on the south side of the A51 road, up to 4.3 m of loose brown sand and stones are recorded. North of the Shropshire Union Canal south-west of Southley Farm [578 593], and in the railway cutting 135 m due south of the farm, up to 9 m of loose brown sand and stones were noted. There is a seepage line at the base of the sands where they are thought to rest on an underlying till.

Farther east, and north of the A51 through Alpraham village, a mass of sand, disposed in drumlin-shaped mounds, extends for over a kilometre north-east of Highwayside [586 597] and is probably continuous underneath the clay cover with the Tilstone Fearnall sands. Many minor exposures of loose sand and gravel were recorded. Sand was dug from an old pit about 2.5 m deep immediately east of Hilbre House [587 597].

Isolated diggings in glacial sands, also probably extensions of the Alpraham mass, include one [587 600] 350 m NNE of Hilbre House, where 3 m of brown clay till overlie loose sand and gravel, and another [578 605] immediately east of Rookery Farm where 2 m of loose brown sand, silty in the upper part, form a small mound capped by brown sandy clay till.

### Bunbury-Bunbury Heath area

Immediately south-east of Bunbury Mill Pond [573 580] glacial sands form a small mound in which 2 m of loose brown sand with stones are exposed at an old pit. Small patches of sand beneath sandy till occur to the south and east of Bunbury village, and it is probable that sand forms the core of the drumlin-shaped mound which extends NNW from Bunbury village to join exposed sands, part of which form a mound 320 to 500 m SE of Priestland Farm [560 587]. Here up to 4.6 m of loose brown medium- to coarse-grained sand with a few pebbles are seen in the banks of a small stream. There are exposures of sand in the banks of the river Gowy south of Bunbury Heath, overlain and, locally, apparently underlain by clay. Glacial sand probably extends south-eastwards towards Haughton and north-westwards to Beeston village under a thin cover of sandy till. Sand probably forms the cores of the two ridges south-east of Beeston village, trending NW-SE.

### Spurstow-Haughton

A low mound south of the Gowy at Bunbury Heath is cored by sand; a pit on its southern flank some 450 m NNE of Haycroft Farm [554 572] showed up to 6 m of loose sand and stones. In a similar north-west-trending mound immediately south of Haycroft, 2 m of loose medium-grained brown sand beneath 46 cm of brown stony clay were seen at an old pit.

Glacial sand forms the core of a drumlin-shaped mound some 100

to 300 m west of Spurstow Hall [569 565] and the two outcrops are probably continuous beneath the clay, with the sand extending from Spurstow Hall to Spurstow School [574 561] and being best exposed in old pits south of Spurstow Hall fishpond where up to 3 m of loose sand are recorded. A well at Spurstow Hall proved 3 m of brown till on 6 m of loose water-bearing sand. West of Haughton Hall an old pit [582 564], at the southern end of a sand-cored ridge extending north-westward for nearly a kilometre, exposed 2.5 m of coarse brown sand and gravel.                                                                                                     E.G.P

## Laminated clay

A common feature in the occurrences of laminated clay detailed below is that each is part of a glacigenic sequence, apparently both overlain and underlain by till or by glacial outwash deposits. The clays represent quiet conditions of deposition in temporary lakes, probably dammed by ice or ice-cored moraine, in contrast to the vigorous regime of glacier drainage that carried in the sands and gravels on which they rest. The bedding distortion noted in the laminated clay at Ash Brook and in the Beeston Castle sandpit is of the order of magnitude that might be expected if a layer of unconsolidated, and extremely weak, material was overridden by flow till generating both loading and shearing stresses.

At Tattenhall Road Brick Works pit [495 607] about 1 m of brown stoneless laminated lake clay underlies 6 m of stiff brown boulder clay. This is the only outcrop of laminated clay large enough to be shown on the 1:50 000 map. The base of the laminated clay is not seen but, only 150 m ESE of the pit, a borehole [495 606] 50 m north-west of Crow's Nest Bridge records: clay 14.5 m; on sand, gravel and marl to 27 m; on red sandstone. Two adjacent boreholes proved 13.5 and 13 m of clay respectively. If a substantial part of the clay recorded in the boreholes is laminated clay, then the deposit may underlie a considerable area, particularly the low-lying ground to the west and south.

At Beeston Castle sandpit [554 595] up to 3.5 m of brown stoneless laminated lake clay rest upon glacial sands, and are in turn overlain by boulder clay (see p.77). The laminated clay is violently contorted under the boulder clay.

On the steep south bank of Ash Brook, some 300 m west of its confluence with the river Weaver, the following sequence is exposed [6660 6307]:

|  | *Thickness* |
|---|---|
|  | m |
| Red till | at least 1.00 |
| Finely varved clay, slickensided and distorted | 0.61 |
| Sand and gravel, abundant flints | 0.46 |
| Sand and silty clay, interbanded | 0.46 |
| Sand, light orange | 0.91 |

From nearby exposures in the stream banks it is thought that bedrock is not far below the surface at this point, and that the glacial sequence is comparatively thin. Nevertheless, the position of the varved clay between an upper till and underlying sands and gravels is analogous to the position of similar clays elsewhere within the district.

Varved clay of unknown thickness was noted in the floor of a sand pit [6024 6993] at Dalefords, near Sandiway. About 6 m of sand overlie the clay, and it is thought that there is also a thick sand sequence below it. No natural distortion of bedding was noted in this clay bed, although there had been disturbance from the action of vehicles in the quarry. The clay is thought to extend over a considerable area around the pit, with consequences on the movement of groundwater in the sands (see pp.73 – 74).

## Late glacial sands

The only deposit of glacial sand and gravel that is probably of later origin than the great mass of material so classified is an extensive spread around Spurstow and this is not shown separately on the 1:50 000 map. Another very small patch formerly capped the high bank above Beeston Castle sandpit.

At Spurstow fine-grained yellow to brown stony sand, very silty in places and about 2 m thick, extends from a field west of the Methodist Chapel [556 570] south-eastwards for about 1.5 km, with a maximum width of over half a kilometre. Its southern limit is defined by Badcock's Lane, north of Fields Farm [563 555]. The deposit forms good sandy loam soil, locally prized for arable farming. It was probably laid down in water standing not far above 80 m OD.

At Beeston Castle sandpit [554 595] the section visible in 1954 (see p.77) showed, above the boulder clay near the top of the quarry face, 137 cm of loose brown sand with few stones resting on 15 cm of gravel composed of pebbles derived from the Pebble Beds, and rounded glacial erratics. A hard black-stained 'foxbench' was visible in places at this horizon, its presence being due to iron cementation of the gravel bed. The deposit lay at about 61 m OD and was probably laid down in water standing not far above this level. EGP

## Fluvio-glacial gravel

East of the southern part of the Peckforton Hills a large area of sand, silt and gravel, mapped as fluvio-glacial gravel, extends south of Peckforton village and around Peckforton Moss [545 560], where up to 2 m of the deposit can in places be seen in drainage ditches. The material is regarded as true morainic outwash fan which grades into the south-west slopes of the Peckforton Hall Moraine (see p. 71).

In two places, at Peckforton Gap [528 555] and farther north [536 566] to the west of Peckforton village, it extends westwards up glacial drainage channels cut into the eastern slopes of the Peckforton Hills. The formation of these deposits in the ice-bound extremities of a lake-system, the main bodies of which lay to the south, is discussed in Poole and Whiteman (1966, pp.71-75).

East of the northern part of the Peckforton Hills many small exposures of sandy silt and gravel, mapped as fluvio-glacial gravel, occur along both sides of the Peckforton to Beeston road, and along glacial drainage channels south-west and north-west of the entrance gate and lodge [538 576] of Peckforton Castle. Numerous exposures of sand, sandy silt and gravel also occur in the former glacial drainage channel now occupied by the headwater brook of the river Gowy (an evident misfit stream hereabouts). The best of these is to be seen [542 572] by the Gowy brook some 400 m south of where it flows into Peckforton Mere; here 1 m of sandy silt rests on 1.5 m of loose yellow-brown fairly coarse sand. These deposits are believed to date from the period when the retreating ice-front halted along the line of the Beeston Castle Moraine (see p.71).

North of the Peckforton Hills, in the gap between Beeston and Peckforton castles, and in the area between Brook Farm [524 591] and Woodhouse Farm [520 580], the present-day small streams have cut down through patchy deposits of loose yellow and brown sand and gravel, brown clay and silt. Though mapped as fluvio-glacial sand and gravel it seem likely that some of this material could be classed as glacial sand and gravel, but the amount of resorting, downwash from the hills, and deposition from the melting ice that have taken place make separation of glacial from fluvio-glacial material impossible.

In a small stream valley [532 586], 275 m NW of Moathouse Farm, up to 4.5 m of loose yellow sand and gravel occur. Downstream these deposits become finer-grained and more silty. An old sandpit [539 587] west of Beeston village is now grassed over but, in a small exposure [538 589] on the north side of the road, 130 m east of Castleside Farm, up to 1 m of loose waterlain yellow-brown sand and gravel overlies 2 m of Upper Mottled Sandstone. West of Castleside Farm, on the southern banks of the small stream

issuing from Beeston Spa [535 590], numerous rabbit burrows have been dug into loose sand and gravel. In a ditch [528 587] along the north-east side of Gregory's Wood, up to 2 m of sandy clay with gravel are seen, and rabbit burrows in the middle of the wood have thrown up loose yellow sand. Up to 1 m of soft sand and gravel is visible in ditches in the immediate vicinity of Woodhouse Farm, and 2 m of loose sand and gravel was seen in a silage pit by the roadside about 100 m WNW of the farm.

The fluvio-glacial gravel deposits depicted on 1:50 000 map along the Weaver and Dane valleys are discussed under River Terraces (p.80). EGP

## Glacial drainage channels

In the north, two well-defined glacial drainage channels are cut into the eastern slopes of the hills in the Newton-New Pale area. Newton Hollow, a broad, gently graded dry valley which courses south-eastwards from a position [521 748] north-east of Rileybank, was probably cut by glacial meltwater. It channels the watershed just above the 122 m contour, falls gently south-eastwards and loses definition about 1.5 km away [534 740] just east of Ravenslow Farm. A much more sharply-defined channel starts on the ridge [521 731] south-east of Claim Farm at a little over 122 m OD and falls quickly south-eastwards along a sinuous course. Near New Pale Lodge [525 727] it is joined on its north side by a deep dry ravine, presumably a run-off channel. Farther south-east the main channel becomes indefinite where its course [528 724] lies a little below 91 m OD east of New Pale Farm.

In the Kelsall-Delamere area, a well-defined channel starts [533 687] on the watershed at the north-east end of Kelsall, where it floors the gap used by the A54. It falls gently eastwards, and is deeper and more gorge-like west of Organsdale House. It becomes vague [546 682] just below 91 m OD south of Organsdale House, where it acquires the shallow stream known as Hindswell Gutter. On the south side of Hindswell Gutter, two run-off channels are deeply incised into the drift-free Helsby Sandstone. The western channel has craggy sandstone banks 6 to 9 m high along much of its course; the deepest part is known as Urchin's Kitchen [541 679]. The eastern channel is somewhat deeper but its sides are less precipitous. It starts at Dungeon Bank [544 675], courses generally north-eastwards where it is known as Dog Fall [547 678], beyond which it fades quickly into a vague depression.

A ravine, which is interpreted as a glacial run-off channel, deeply scores the west-facing escarpment overlooking Willington Corner. It starts [538 671] south-east of Roughlow Farm at about 130 m OD and runs generally south-westwards, dying away rather abruptly where it debouches [534 669] at about 80 m OD on the drift-covered ground north-east of the crossroads at Willington Corner. Between Utkinton Hall and Utkinton Mill Pond [563 649] the watershed of the mid-Cheshire hills is traversed from west to east by a deep dry valley. For about 300 to 400 m east of Utkinton Hall the depression seems to be rock-floored, and it is possible that this conspicuous feature was partly cut by glacial drainage.

West of the Peckforton Hills there are several glacial drainage channels, the general positions and functions of which were noted by Poole and Whiteman (1966, pp. 71, 75, 87–88). The most spectacular of these is Peckforton Gap [528 555], which trends in a south-easterly direction to just beyond the southern limit of the district. This channel, and others near Peckforton village and near the entrance to Peckforton Castle, have been noted above (p. 71).

### The Deva Spillway

This fine glacial drainage channel, here named the Deva Spillway, is a major topographical feature of the north-western part of the district. It has been eroded through boulder clay and glacial sand and gravel, and for much of its length it is cut into solid Pebble

Beds. It can, therefore, be dated with some certainty as late-glacial or post-glacial in age. The highest part of the channel floor at its in-take end [420 725] is by Croughton village where it passes from the Mersey into the Dee catchment at about 12 m OD. It courses south-westwards and westwards for nearly 3 km to Backford Bridge [400 710], beyond which it swings south for another 3.5 km, much of its channel forming the extensive grounds of the Deva Hospital. Its mouth [397 674] is near Blacon House Farm in the north-west en-virons of Chester, and the discharge level is that of the estuarine alluvium of the river Dee at about 5 m OD. The original discharge level was almost certainly much lower, for its lower reaches are much silted-up with recent alluvium.

A drainage channel of this size, gradient and location has clearly played a major part in the deglaciation history of the Mersey-Dee drainage system and its role is discussed below.　　　　　JRE

## River terraces

### River Dee

From the southern margin of the district to the river bend [414 619] just south of Eccleston Ferry, there is a well-defined terrace, map-ped as Second Terrace, that flanks several kilometres of the floodplain alluvium of the Dee at a height of 4.5 to 6 m above the floodplain. In Eaton Park it lies west of the floodplain, but farther south it is mainly to the east. The greater part of Aldford village is built on this terrace and, south of Aldford, it is almost continuous along the east side of the river, or its floodplain, to the southern margin of the district. East of Eaton Hall the terrace is channeled by a contemporaneous abandoned river course for a distance of nearly a kilometre. The terrace deposit consists of sandy loam and gravel which has been considerably dissected in places. The tract between Aldford and Farndon (Sheet 122) forms a well-drained gravelly soil, particularly suited to the growing of strawberries. In the extreme south-west of the district, where the floodplain of the river Alun merges with that of the Dee, there are traces of a similar terrace, mainly along the south flank of the alluvium of the Alun.

There are four strips of terrace, mapped as First Terrace, bet-ween Eccleston Ferry [415 622] and the extreme south-west of the district. They flank the floodplain alluvium of the river Dee and lie at a height of 1.5 to 3 m above it. A small area of the terrace lying east of the river immediately south of Eccleston Ferry carries only a thin gravelly spread of re-worked boulder clay; there is another small strip [421 601] near the mouth of Aldford Brook.

The best development of the First Terrace is east of the river Dee at Aldford, where a strip of terrace up to 65 m wide extends for a kilometre or so between the floodplain alluvium and the Second Terrace. It lies only about 1.5 m above the alluvium and is compos-ed of sandy silt and stones. Another strip of this terrace is seen [390 558] west of the river Dee, and south of Ithel's Bridge over the river Alun, where it is about 830 m long and up to 130 m wide, at a height of 1.5 to 3 m above the floodplain alluvium. The terrace here contains much more clay-grade material, probably re-worked boulder clay.

### River Gowy

Two patches of terrace on the south side of the upper reaches of the Gowy have been mapped as Second Terrace, but they are too small to be recorded on the published map. The larger is a triangular patch of re-sorted glacial sand immediately west of Beeston Castle Inn [553 597]. The smaller lies south-east of the aqueduct [509 603] carrying the Shropshire Union Canal across the river, and consists of stony silt which is only slightly coarser than the First Terrace and alluvium of this tract.

A terrace mapped as First Terrace, lying at a height of 1.5 to 3 m above the alluvium, is well developed along the river Gowy and its tributaries from Tilstone Mill [566 594] to Higher Huxley Mill [506 608]. Farther downstream there is an isolated patch [494 626] nor-thwest of Lower Huxley Mill. The terrace consists of sandy silt and sand with few stones, and is similar in composition to the alluvium.
　　　　　EGP

The only other tributary of the river Gowy along which a well-defined terrace is preserved is the brook north of Duddon Common, part of the course of which is artificially diverted to flow through Duddon Mill. Here, a terrace mapped as First Terrace, lying about 1.5 m above the alluvium, extends from a point [525 655] north-east of Duddon Common to about half a kilometre north-west of Dud-don Mill.　　　　　JRE

### River Weaver

Around the town of Nantwich, in the district to the south, a Second Terrace is mapped at about 7.5 m above the normal level of the river Weaver. Its outer edge is difficult to determine because the terrace is nearly at country level (Poole and Whiteman, 1966, pp. 91–92): it is possible that a fairly extensive shallow lake formed at this level to the west and east of Nantwich. The terrace edge becomes more pronounced downstream from Nantwich as the river falls with respect to the country level, and wide well-marked terrace flats, occupied by rather thin spreads of loamy stony sand, flank the Weaver where it enters the present district north of Nantwich. The overall original width of the terrace system increases from about 300 m east of Rookery Hall [660 560] to nearly a kilometre north of Church Minshull and near the confluence [669 631] with Ash Brook. Northwards the terrace rapidly degenerates. Between the Ash Brook confluence and Top Flash the terrace is represented only by three loamy sand-covered benches on the clay valley sides, and these diminish in size downstream. Below Top Flash there is no vestige of a terrace at this level for 8 km downstream, with the ex-ception of a narrow bench on the left bank near Vale Royal, as far as Hartford Bridge [647 713]. It may be significant that the point along the river where the terrace system vanishes is also where the river enters an area of extensive subsidence, with many manifesta-tions of ground instability even at the present day. From Hartford Bridge to the Dane confluence the Weaver is flanked on one side or the other by wide sandy terrace flats, still about 7 m above the nor-mal river level.

A wide meander of the river Weaver between Weaverham and Winnington was flanked by a strip of high terrace up to 600 m wide. Much of this area is now occupied, as is the adjacent alluvial flat that was once the floodplain, by settling tanks for chemical waste, and the geological features are obliterated.

The Second Terrace is shown as a fluvio-glacial deposit, follow-ing the practice on the Stockport (98) and Runcorn (97) sheets to the north. It had been noted that the valleys of all the rivers that drained to the Mersey estuary contained extensive sheets of sand and gravel which formed a high terrace, in places 10 m above the present river level. It appeared that the formation of these terraces was terminated by an event which suddenly lowered the base-level to which the rivers were being graded, and that it happened to all the rivers simultaneously, for their high terraces are all at grade with one another—the Bollin with the Mersey, the Mersey with the Weaver, and so on. The later terraces form a group, including the modern alluvium, that are confined to the immediate vicinity of the river. No such grouping of terrace levels is noted in the lower reaches of rivers flowing to the Dee estuary. The explanation adopted in the Stockport memoir (Taylor and others, 1963, p.110) was that of Dr J. R. Earp, that, when the terraces were forming ice still blocked the Mersey estuary, so that all the drainage of the Mersey and its tributaries was diverted along the Deva Spillway to the Dee valley (see above). The high terraces were therefore, graded to the intake level of that channel. The levels in the Deva Spillway are in agreement with this hypothesis, and the channel was undoubtedly a major water-course in late-glacial times. An alternative explanation for the terraces (Evans and others, 1968, p.250) is that they represent a high relative sea- level at a time when the rise in the level of the oceans through ice-melting had outstripped the recovery of the land from its isostatic depression under ice load. This implies that they are younger than does the theory linking them to an ice-front.

The high Weaver terrace, like those of the Mersey and its tributaries, was regarded, at the time of survey, as 'fluvio- glacial', because it was thought that when it was being formed ice was still exerting an influence on the drainage. Any other explanation removes the justification for this classification.

The only terrace level along the Weaver in this district that is intermediate between the Second Terrace and the modern alluvium is represented by two small sand-covered benches mapped as First Terrace [664 614, 673 629] on the left bank of the river, only a metre or so above the floodplain, between Church Minshull and the confluence of the Weaver with Ash Brook.                    BJT

East of Moss Farm [578 564], Haughton Moss, on the northern side of the brook that flows via Barbridge to the river Weaver, an alluvial tract mapped as First Terrace, with a peaty or black silty surface, extends over about 20 hectares. It passes westwards into the peat of Haughton Moss, and it is possible that this area was once a lake which has been drained and left as a terrace by the down-cutting of the tributary.                    EGP

### River Dane

Most of the river Dane is in the Macclesfield (110) district to the east, and the Dane terraces were described by Evans and others (1968, pp. 216–217, 237–238). Only the lowest 5 km of the river, from Bostock to the Weaver confluence, are in this district.

Four patches of the highest terrace remain: a sandy bench [673 698] of about 3.5 hectares 500 m NE of Bostock Green: a flat [675 705] of about 8 hectares south of Shipbrookhill, where numerous rabbit burrows have been dug in loamy and silty stony sand; and two sand-covered flat areas of 1.2 and 4.5 hectares respectively separated by Eldersbriar Brook due east of Davenham. All these terrace remnants are about 6 m above river level. They are at grade with the high (Second) terrace of the Weaver.

There is a lower terrace about 1.5 m above the floodplain alluvium, of which patches remain due east of Davenham, adjacent to the areas of high terrace described above, and on the left side of the floodplain [666 728] north-east of Leftwich Green.                    BJT

## Alluvium and estuarine alluvium

### River Dee

Between its confluence with the river Alun [399 561] in the extreme south-west of the district and the rock-cut gorge at Chester, the river Dee meanders through broad tracts of floodplain alluvium within which several abandoned meanders can be recognised. The best of these is the Serpentine Lake [415 600] in Eaton Park. The floodplain is broadest at Trefalun Meadows, around the confluence, where it covers some 4 km², and although the alluvium here reaches a height of up to 3 m above normal river level it is often flooded in the winter. A shallow well at Meadowhouse Farm [390 570] at about 6.5 m OD penetrated nearly 6 m of alluvial sands and gravels. Downstream, in the Aldford-Eaton reaches, the floodplain averages about 500 m in width and stands about 1.5 m above the normal river level. At Aldford the alluvium consists of fairly fine sandy brown silt, but the deposit becomes coarser upstream; this may be because the local source of coarse material is the Pebble Beds ridge which is breached by the river at Farndon (Sheet 122).

Between Eccleston Ferry and Heronbridge the river lies close against the eastern flank of the Pebble Beds ridge and the floodplain is only 100 to 200 m wide. At Heronbridge it is confined by Pebble Beds on both flanks, but between Heronbridge and Earl's Eye it broadens out again to an average width of about 500 m. Below the weir at Chester the alluvium of the Roodee (Chester Racecourse) and Bache Brook areas merges westwards into the great spread of estuarine alluvium that extends to Sealand and the present-day estuary. The estuarine alluvium consists of shelly sand and silt.

Of the eastern tributaries of the Dee, Aldford Brook and its tributaries, the chief of which are Plowley Brook, Coddington Brook, Mere Brook and Golborne Brook, meander through almost continuous tracts of alluvium which, along the main watercourse, are commonly up to 250 m wide. On these wide expanses the alluvium stands one to two metres above normal stream level. Narrower strips extend a long way up each minor valley, and in some cases almost up to the source of the stream. All this alluvium is generally fine in grain and composed of re-sorted boulder clay and silty sand with scattered stones.

Farther north, Henlake Brook and its upper reach Powsey Brook have alluvial flats more than 50 m wide, and there are similar tracts along Caldy Brook and its tributaries in the south-eastern suburbs of Chester. North of Chester the broad glacial drainage channel named the 'Deva Spillway' is floored for nearly 6 km by an alluvial flat 50 to 150 m wide. This alluvium is mostly dark grey, locally peaty, clay or loam, and the peaty character is very pronounced in the tract south of the Shropshire Union Canal between Caughall Bridge [413 711] and Backford Bridge [400 710].

### River Gowy

Above Tilstone Bank, the river Gowy headstream and its main tributary that drains Spurstow, Woodworth Green and Calveley flow on marshy alluvial tracts up to 50 m wide, much of the alluvium lying less than 1 m above normal stream level. The collecting ground around Peckforton and Spurstow has many irregular areas of peaty alluvium up to 500 m across. The largest of these lie between Peckforton Moss and Peckforton Mere, between Beeston Moss and Haycroft [554 572], and east of Radleywood [559 564]. Below Tilstone Bank down to Higher Huxley Mill the alluvium of the Gowy is of less extent than the First Terrace, and lies from one to two metres above normal river level. Below Higher Huxley Mill down to its confluence with Milton Brook [458 682] the Gowy, for much of its course artificially straightened and banked, occupies a continuous floodplain that reaches a maximum width of about 500 m. The alluvium makes peaty marshland south-west of Brereton Park Farm [498 629], and there are patches and mounds of thin peat south-east of Walk Mill [484 643]. From above Ford Bridge [482 648] down to the mediaeval bridges at Hockenhull Platts [476 657] the alluvium is sandy and locally gravelly, probably because of the proximity of outcrops of glacial sand around Stapleford Hall. Eastern tributaries of this part of the Gowy are Southley Brook, Waterless Brook and the brook east of Walk Mill, all of which have alluvial tracts for a kilometre or so above their confluences. Western tributaries are Guylane Brook and the brook west of Stanford Mill [468 668] which have similar short alluvial tracts, but the higher reaches of Guylane Brook are also on a narrow alluvial floodplain.

Between the Milton Brook confluence and Bridge Trafford the river Gowy makes a broad alluvial floodplain, the southern part of which encompasses the 'island' of Plemstall, and the northern part includes the alluvium of Barrow Brook that reaches a maximum width of about a kilometre. The alluvium consists of grey or dark grey clay or loam, the surface of which is locally peaty, and around the northern part of the Plemstall 'island' there are several small areas of thick peat. Below Bridge Trafford, where the floodplain is between one and two kilometres wide, much of the surface is peat-covered, but west of Thornton- le-Moors there is a large area mapped as alluvium where brownish clay, up to a metre in thickness, rests on either peat or bleached silty clay or boulder clay. Undoubted estuarine silt was not seen, although the floodplain to the north has been mapped as Estuarine Alluvium in adjacent parts of the Runcorn (97) district.

From Duddon Common to its confluence with the river Gowy, Milton Brook flows on an alluvial floodplain from 100 to 150 m wide, reaching its greatest width below Barrow Hill [471 684]. Between Duddon Mill [513 658] and Oscroft Bridge [501 666] the lower part of the alluvium is very sandy; the upper part is more silty and includes much peat. Below Tarvin only grey or dark grey, locally peaty, clay or loam were seen. The only other important tributary of the river Gowy is Barrow Brook, formed by the union, east of Little Barrow, of Salter's Brook and Ashton Brook. Salter's Brook has a fairly narrow alluvial floodplain that extends from just above Saltersford Bridge [499 679] down to Little Barrow. The alluvium is mostly grey or dark grey clay or loam. Ashton Brook, although mostly occupying a steep-sided valley, has a well-defined alluvial floodplain from its culvert beneath the railway at Mouldsworth down to Little Barrow. For much of its course the brook occupies an artificial channel and the sandy levees of its original bed are visible in places. The alluvium is mostly very sandy owing to the extensive outcrops of glacial sand through which its valley is cut. Small areas of peat occur south of Mill Wood [495 700] and again [490 700] by the mouth of a small unnamed tributary from the north. JRE

### River Weaver

The present course of the Weaver, from its entry into the district near Rookery Hall [660 560] to its confluence with the river Dane at Northwich was determined in early post-glacial times by minor irregularities on the surface of the thick glacial deposits. After the abandonment of the high terrace level, described above, there was a rapid downcutting. Some of the meanders in the higher reaches around Church Minshull, that were clearly formed during the high terrace period, were incised into the underlying boulder clay and bedrock, and since that time there has apparently been very little change in the general shape of these meanders. The river bed has remained within a sinuous strip which varies generally from 80 to 150 m in width, rarely widening out to more than 200 m. It is probable that the lower reaches of the Weaver, like the lower parts of the rest of the Mersey-Weaver system, were graded to a very low sea level following the high terrace formation, and that the present alluvium represents an aggradation which occurred as the sea level rose to its present height (the 'Flandrian Transgression'). This is borne out by the fact that there are no exposures in the river bed of material underlying the modern alluvium, except at the extreme edges of the floodplain. Compared to the rather gravelly sand of the Second Terrace, the Weaver floodplain alluvium is much finer, consisting of light brownish grey sandy or silty clay. Pebbles are less abundant than in the terrace deposits.

Upstream from Winsford, much of the Weaver alluvium is submerged beneath the waters of the three large subsidence meres formed where the river crosses an area of instability due to salt-solution. The rate of subsidence has greatly exceeded the rate of deposition in these stretches of the river, and fresh subsidence scars on the hillsides flanking the 'flashes' indicate that ground movement has by no means yet ceased.

Below Winsford the Weaver was straightened and canalised in the eighteenth century for navigation by vessels engaged in the salt trade: this was an important factor in Winsford's prosperity. Locks were installed to compensate for the shortening of the river line, and those at Vale Royal and Northwich lie within the district. The new cut has left several abandoned small meanders, but this part of the river never developed widely-swinging meanders like those around Church Minshull or on the river Dane, its course being in a rather narrow steep-sided valley below Winsford Bridge.

Below the confluence with the river Dane the Weaver swings in wide meanders through Northwich and Anderton. Parts of its course have been extensively modified both for navigation and as a consequence of serious — in places catastrophic — salt solution subsidence in the town. The joint alluvial floodplain of the Dane and Weaver is 800 m wide in the central parts of Northwich. West of Winnington the modern alluvium follows a wide meander that is incised through boulder clay into the underlying Mercia Mudstone bedrock. Part of the floodplain south-west of Winnington is occupied by large settling tanks containing waste lime from nearby chemical works.

Ash Brook and its tributaries Bankside Brook, Wettenhall Brook, Darley Brook and Chesterlane Brook, all have uncomplicated alluvial patterns. There are no terraces, and the alluvial strip is rarely more than 50 m wide. As in the case of the Weaver, exposures of pre-alluvial beds within the width of the floodplain are confined to the extreme edge, particularly on the outsides of river bends. This suggests that aggradation during the Flandrian reached well up into the headwaters of the river system.

### River Dane

The Dane enters the district about a kilometre north-east of Bostock Green. The floodplain is wider than that of the Weaver, averaging 400 m in the 5 km of the Dane valley that lie within the district. Near Lower Shurlach, an eastern suburb of Northwich, the river swings sharply from a general northerly to a westerly course before joining the Weaver. In this short stretch of the river some meanders have been cut off by artificial channels, and there are signs that this effective shortening of the river, and the consequent steepening of its fall, are having an effect of the banks highers up the river. As has been pointed out in the case of the Weaver, exposures of pre-alluvium deposits along the river are confined to the edges of the floodplain, particularly where these coincide with the outsides of meanders. This supports the view that the alluvium represents an aggradation from a lower base level in the regime of the river.

BJT

### Peat

Small areas of peaty soil and patches of thin peat, too thin to map, are common on the alluvial floodplains of the river Gowy and its tributaries, and some occur along the eastern tributaries of the river Dee. Thicker peat occurs on the Gowy floodplain around Plemstall. This village occupies a boulder clay 'island' surrounded by floodplain deposits, and at one time it must have been almost 'moated' by waterlogged peat-marsh. The peat mapped between Plemstall Bridge [445 701] and St Peter's Church is more than 1.5 m thick, and in several places shell marl with a rich molluscan fauna has been dug out from beneath it (see p. 83). Another spread of thick peat occurs just west of the A56 at Bridge Trafford, and a small area near the middle [447 710] is slightly raised. The largest expanse hereabouts occupies the floodplain between Picton and Thornton-le-Moors, covering an area nearly 3 km in length and up to 1 km wide. Much of this peat exceeds a metre in thickness and forms the soil, but on either side of the river the soil includes 15 cm or so of dark grey alluvial clay, probably brought in by flood-water after the area was drained.

Most of the numerous enclosed hollows that modify the otherwise rather even surface of the Delamere Sand are probably kettle holes, resulting from the melting of patches of ice buried beneath the outwash deposits, and thick peat occupies many of these hollows. The larger depressions were formerly meres, for they lay below the local water-table in the sands, but most were drained and planted many years ago. There are numerous peat-filled hollows of various shapes and sizes in the thickly-wooded northern area of the Delamere Forest up to 2 km north of the railway line between the Youth

Hostel [530 713] and Hondslough Farm [540 725] in the west, and Gallowsclough Hill [571 711] and Breech Moss, Norley [566 723] in the east. Some of the larger mosses still retain their pre-drainage names such as Flaxmere [556 723], Linmer Moss [547 707] and, the largest, Blakemere Moss [552 711], which covers an area of 0.5 km². Hatchmere [553 722] is still partly occupied by water, but a strip of peat extends north-westwards from the lake and appears to rest on a clay which is interpreted as an intermediate flow till in the Delamere Sand (see p. 74). The peat is said to have yielded a Zone II flora (Boulton and Worsley 1965, p. 706), and this fixes a younger age limit for the till-sand sequence in the Hatchmere area at about 10 000 years BP (see p. 65). South of the railway, from near Delamere Station to a short distance south of the A556 at the Abbey Arms crossroads, there is a belt of peat-mosses running parallel to the rising ground of the mid-Cheshire ridge to the west. Among the largest of these Relick's Moss [560 700] is disrupted and partly flooded by sand quarrying operations. The peat deposit in the south-eastern angle of the Abbey Arms crossroads was proved to be 6 m thick at the Fourways Cafe Borehole [5654 6861].

Oakmere, and the peat-filled hollows to the north-west and north-east of it, as well as a string of irregularly shaped peat basins to north and south of the A556 [eg. 570 688, 568 594] were probably formed by salt-solution subsidence rather than ice-wasting. Active subsidence was for many years suspected at Oakmere, and salt was finally proved in that area first by the Bowyer's Waste Borehole, and then by the Oakmere Borehole (see Appendix 1). The East Delamere Fault, bounding the saltfield on the west, is drawn west of the Bowyer's Waste and Oakmere boreholes and east of the Fourways Cafe Borehole which proved sandstone in the Sherwood Sandstone Group beneath the drift.

East of the A49 in the Abbot's Moss Wood area, there are poorly drained peat-filled hollows linked with the marshy upper reach of the valley leading eastwards past Dalefords to Petty Pool. One of the largest of these is known as Shemmy Moss [595 689]. A nearby peat-moss [601 688] is also more or less permanently waterlogged. Farther north there are two areas of peat in Barry's Wood [597 701].

Around Sandiway the main occurrences of peat are Cartledge Moss [602 713], Littledate's Gorse [615 715] and an area [620 708] just west of Sandiway Farm. Farther east again there is a large expanse of peat-marsh in the valley of Pettypool Brook that extends from imediately west of Whitegate village for more than 500 m to the west.                                                    BJT, JRE

## Shell marl

### Plemstall area

Grey silt with abundant freshwater shells, dug from beneath the area of thick peat on the south side of the river Gowy immediately west of Plemstall Church [457 701] in 1948–9, was investigated by Nora F. McMillan (1958). The deposit is of unusual interest in that it yielded many valves of *Dreissena polymorpha* (Pallas), a species formerly recorded in the fossil state only from shell marls at Clifton Hampden near Abingdon and at the sites of Ugg Mere and Whit-

tlesey Mere in the Fens. The mollusca indicate that, prior to the formation of the peat, the area was almost certainly occupied by a shallow mere or lake-like expansion of the stream, and the presence of anthers of *Alnus* in the silt is indicative of the presence of alder-carr in close proximity.

The gastropod fauna includes: *Lymnaea auricularia* (Linne), *L. palustris* (Muller), *L. peregra* (Muller), *L. stagnalis* (Linne), *L. truncatula* (Muller), *Planorbis planorbis* (Linne), *Gyraulus albus* (Muller), *Bathyomphalus contortus* (Linne), *Anisus vortex* (Linne), *Acroloxus lacustris* (Linne), *Bithynia tentaculata* (Linne), *Valvata cristata* Muller, and *V. piscinalis* (Muller).

The bivalve fauna includes: *Dreissena polymorpha* (Pallas), *Pisidium amnicum* (Muller), *P. casertatum* (Poli), *P. henslowanum* (Sheppard), *P. hibernicum* Westerlund, *P. milium* Held., *P. nitidum* Jenyns, *P. obtusale* (Lamarck), *P. pulchellum* Jenyns, *P. subtruncatum* Malm., *Sphaerium corneum* (Linne), and *S. lacustre* (Muller). There were also a number of ostracod valves including: *Candona candida* (Muller), *C. compressa* (Koch), *C. albicans* (Brady), *Cyclocypris ovum* (Jurine), *Cypris exculpta* (Fischer), *Herpetocypris reptans* (Baird) and *Lymnocythere inopinata* (Baird).                                                    JRE

### Spurstow area

In a small ditch 410 to 500 m due east of Radleywood [559 564] the following section was exposed: black peat with abundant woody fragments, branches and buried tree roots, 61 cm; on shell marl, white, greasy, with abundant gastropods and some bivalves, and with abundant plant and wood fragments, roots, etc., throughout, 41 cm. Only 41 cm of this shell marl was seen at the bottom of the ditch, but the farmer at Radleywood believed that it extended downwards for at least a further 1.8 m. The flat peaty and alluvial tract here is over 24 hectares in extent and was originally a lake, probably a relict late-glacial mere, which has been drained by a cutting through the boulder clay at its eastern end.                          EGP

## Older blown sand

Within an area of about 0.5 km² around Thornton Green [444 735] the boulder clay is covered by a thin spread of light grey sand that makes a characteristic grey sandy soil. The deposit extends from the vicinity of Cross House Farm [450 733] westwards to the peat flats of the Gowy floodplain, and it probably extends for short distances beneath the peat. No good sections were visible at the time of the survey but the presence of a number of shallow clay pits in fields where the soil is very sandy show that it is not a deep-seated deposit like much of the glacial sand, and probably does not average more than a metre or so in thickness. The sand grains are almost entirely of silica, and the presence of small faceted pebbles (*dreikanter*) embedded in the sand suggests an aeolian origin.

Strahan (1882, p.30) refers to this spread of sand as 'the most southerly outlier' of the Shirdley Hill Sand, the well-known and economically valuable silica sand of south-west Lancashire. As the Thornton Green deposit lies so far from the main spreads of the Shirdley Hill Sand it is thought better to describe it no more specifically than as 'older blown sand'.                          JRE

CHAPTER 7

# Mineral products, groundwater and civil engineering aspects

## ROCK SALT AND BRINE

Salt or sodium chloride (NaCl) occurs both as bedded rock salt, which in its commercial form is the mineral halite together with up to about 5 per cent of impurities such as silt, clay and salts of calcium and magnesium, and as natural or 'wild' brine overlying the salt surface in the subcrop zone. The resource in this district is in the Northwich Halite, a south-easterly dipping sheet of saliferous beds subject to some faulting. The subcrop underlies a broad belt of country from Spurstow to Winsford and Northwich, and the base of the formation reaches a depth of some 560 m below sea level in the south of the district west of the Winsford Fault.

About 105 km² of the district are underlain by the Northwich Halite at its full thickness of 218 m, given by a composite section from the Worleston, Over and Southley Common boreholes (see Figure 13). The content of clay and silt in mudstone bands and inclusions in the salt is the equivalent of 53 m thickness (p.43), or about 24 per cent of the formation by volume. With a specific gravity of 2.2 there are, therefore, about 38 300 million tonnes of halite in the district, or just under one tenth of the total estimate to exist in the Cheshire-Shropshire basin as a whole (Pugh, 1960, p.279). This calculation ignores the hypothetical feather-edge of the Wilkesley Halite mapped in the Barbridge-Aston juxta Mondrum area in the south-east of the district.

Salt is used as a saturated aqueous brine in the production of chlorine and the alkalis sodium carbonate and caustic soda; as 'white salt' produced by the evaporation of brine in the flavouring and preserving of foodstuffs and in water softening; and as crushed rock-salt on roads for snow and ice clearance. The method chosen for winning salt from the ground depends on the purpose for which it is required. Brine used as a chemical feedstock, and salt for food processing and culinary and table use generally require a purity greater than 99.5 per cent, and this is best achieved by using solution methods, either natural or artificial, to raise the mineral. For uses where up to 5 per cent impurities can be tolerated the preferred method is the direct mining of rock-salt.

Technically there are three possible methods of working the salt; the pumping of 'wild' brine from the 'wet' rockhead in the subcrop zone, the development of artificial solution cavities in the salt beds, and the mining of rock-salt. Pumping 'wild' brine was for many years the chief method of working in the Winsford area, and white salt of various grain sizes was produced by evaporating the brine at different rates in coal-fired open pans. Figure 19 shows the position of brine wells in the town at the beginning of the century, and Figure 16 the subsidence areas for which that abstraction was responsible. All these wells are now abandoned. The catastrophic subsidences in Northwich (p. 46) were due to

uncontrolled pumping of both 'wild' brine and the brine from flooded salt mines. Although there are two companies still (1981) producing salt from natural brine in the Northwich area, the general County Council policy at present is to discourage uncontrolled abstraction, and it is unlikely that new planning permissions will be granted. The vacuum multiple effect system of evaporating brine was pioneered at Winsford in 1905 by the Salt Union, using 'wild' brine feedstock. The vacuum plant now uses brine piped into the area from controlled pumping operations at the Holford Brinefield.

Solution mining, or brine production from artificial solution cavities in the salt beds, was first pioneered in this country by Imperial Chemical Industries, Ltd., at Preesall, Lancashire, in 1892 and later in the Holford area east of Northwich (Evans and others 1968, p.263). The method is described by Notholt and Highley (1973, pp.17–19). Ideally, it requires blocks of relatively unfaulted ground in which adequate thicknesses of sound salt exist both to provide the resource and to allow the development of pressure-tight cavities. It was recognised in the 1952 Cheshire Brine Pumping (Compensation for Subsidence) Act that the risk of subsidence from this method was much less than from 'wild' brine pumping; moreover, if subsidence did occur from the collapse of a cavity it would be on the operator's own land. Another advantage of the method is that the insoluble marl, falls to the bottom of the cavity and is left in the ground. Suitable conditions for operating the Holford system probably exist within the area mapped as the middle division of the Mercia Mudstone in the south-eastern part of the district.

A method of solution-mining of salt developed in the USA, but not generally employed in this country is the hydrofrac method of linking two nearby boreholes in salt beds by applying excess water pressure to one, in order to break through to another, thereby creating a horizontal solution cavity. Control in this technique is probably not so positive as in the Holford method.

Mining of rock-salt in the Northwich area followed accidental discovery of the mineral in an exploratory borehole for coal at Marbury, north of Northwich, in 1670. Graphic accounts of the development of the Northwich mines, their flooding and collapse, and their subsequent use as brine reservoirs causing the devastation of hundreds of acres of ground, are given by Ward (1873, pp.39–63; 1898, pp.274–306, 530–567; 1900, pp.241–264) Dickinson (1882) and Calvert (1915, pp.201–250, 304–374). Meadowbank Mine, Winsford, was originally sunk in 1884 and worked until 1888 when the Salt Union ceased production of rock-salt at Winsford but maintained the mine and shafts in good repair. Production was restarted in 1928, following the flooding and closing of the Adelaide Mine, the last rock-salt mine in Northwich. Since then Meadowbank has been the only rock-salt mine worked in Great Britain.

The salt at present worked at Meadowbank is from the lowest bed of the Northwich Halite. Known as the 'Bottom' or 'Hundred-Foot' salt, the lowest 6 m contains about 95 per cent sodium chloride, being sufficiently pure for such uses as a fertilizer for sugar beet, a flux in aluminium refining and road de-icing. An extensive programme of geophysical survey and exploratory drilling was started in 1956 by ICI Ltd., in the country around the mine with a view to raising the output from around 45 000 tonnes to a possible 300 000 tonnes per annum. Working plans were based on the room and pillar system of mining, with pillars 50 feet (15.24 m) square and rooms 50 feet wide, giving 75 per cent extraction of the rock in the part of the bed mined, and with a working face height of 20 feet (6 m). This potential has now been reached, and the exploratory programme has made possible important revisions to the geological map of the area (see p. 52).

The existing mine is in the subcrop area of the Northwich Halite, with a much diminished, though adequate, thickness of salt above the worked zone. It seems likely that the area mapped as the middle division of the Mercia Mudstone in the south-east part of the district, where the top of the Northwich Halite is 'dry', may contain potential safe sites for rock-salt mining. Since both the Worleston and Wardle boreholes (see p.52) proved that the complete Northwich Halite sequence is present there, horizons other than the 'Bottom Bed' might be suitable for exploitation. It is, of course, possible that faulting of the same intensity as was discovered by the exploratory programme in the Winsford area may exist in this part of the district and for this reason, if for no other, any mining venture would need to be preceded by a thorough exploratory programme.

## SAND AND GRAVEL

### Glacial Sand and Gravel

East of the river Dee and west of the mainly drift-free rising ground of Manley, Tarporley and the Peckforton Hills very wide spreads of Glacial Sand and Gravel intervene between the boulder clay, which forms nine tenths of the surface, and bedrock. The sand shows through where the boulder clay is locally absent, or where steep-sided valleys cut through the overburden, the main outcrops being at or near Wimbolds Trafford, Bridge Trafford, Newton-by-Chester, Hoole, Christleton, Rowton Moor, Aston and Clotton. South of the latitude of Saighton, 6 km SSE of Chester, there are few signs at surface of Glacial Sand and Gravel, the best being near Burwardsley. Sandpits and their sections noted at the time of survey are given on p.72. Few if any sites in these areas will be free from problems of overburden. Glacial sands at a higher level along the edge of the hills between Alvanley and Kelsall, and in the upper reaches of Ashton Brook have been dug in a number of pits (p.72). Much of the sand and gravel proved by boreholes in the western part of the Chester district lies under a thick clay overburden which in most places prohibits economic working. Sections are given on pp.67, 72.

East of the central uplands by far the most important deposit, in terms of both size and ease of working is the Delamere kame-belt, a large sheet of sand and gravel in the triangle Kingsley, Hartford and Little Budworth. Serial sections of the deposits are shown in Figure 24. In the Norley, Cuddington and Sandiway areas there is a patchy overburden of sandy boulder clay, and an overlying till has also been mapped between Whitegate village and the old railway line to the south-west. More or less persistent clay bands occur in places interbedded with the sand, and in the Sandiway area a number of remarkable 'clay intrusions', conical pillars of an underlying till, interfered with the working of the sand. The thickness of the sand over most of the exposed area varies with the surface elevation, and there is probably a gradual rise in the surface of the underlying bedrock towards the sandstone hills to the west. Typical maximum thicknesses are: Sandiway area 61 m with intermittent clay overburden; Newchurch Common 46 m; Nova Scotia [611 686] up to 46 m with intermittent clay overburden; Marton Hall area 30 m. Gravel occurs sporadically in small bands and lenses throughout the kame-belt, but nowhere does it constitute an economically important proportion of the deposit.

The depth of the water-table below surface may be the deciding factor in choosing the method of working. In the northern part of the deposit from Norley to Sandiway, pits can be worked dry with faces up to 10 m high. Farther south, near Nunsmere and Newchurch Common, the standing water-level is in places within 3 m of the surface, and here the method of working has been by floating suction-dredge.

The Marton-Crewe sand-cored esker ridge (see pp.74–75) extends from the south-eastern edge of the Delamere kame-belt at Marton to the western environs of Crewe. Sand is patchily exposed along the crest, but much of the surface is of boulder clay which thickens away from the ridge crest. At the now-abandoned sand pit at Over [635 659], which is a short distance west of the ridge crest, the clay overburden thickened from nothing to an economically limiting 2 m in the 200 m between the east and west sides of the pit. The thickness of the deposit is here about 30 m at its maximum (see Figure 24.7). The water-table was some 16 m below the original surface near the centre of the pit, and a sand face of some 15 m was possible in a dry working. At the time of survey there were further sand pits along the ridge, the most substantial being at Church Minshull [662 614]. The maximum thickness of the deposit is probably maintained as far south as Swanlow Farm [653 635], beyond which it tapers both vertically and laterally. At Church Minshull sand pit [663 614] there were at least 10 m of sand but the overburden increased rapidly westward.

Other sand-cored ridges of similar form and trend occur in the Weaverham, Moulton, Davenham and Bostock Green areas in the north-east of the district, and around Calveley, Barbridge and Cholmondeston in the south-east (see pp.75–76). In all the sand-cored ridges the deposit is mainly sand; gravel is virtually absent. Other deposits of Glacial Sand and Gravel occur in the Tiverton, Alpraham, Beeston, Bunbury and Spurstow area (see p.78). Although pebbles are recorded at many of the exposures, and there are thin bands of gravel in the Beeston Castle sand-pit section, sand is the main deposit and significant gravel deposits are unlikely to be found.

### Fluvio-glacial Gravel

Fluvio-glacial Gravel is mapped west and south of Beeston, and the highest terraces of the rivers Dane and Weaver area also put in this category. The deposit at all these localities rarely exceeds 2 m, and is commonly less; and sand greatly exceeds gravel as a proportion of the deposit. These areas do not, therefore, offer significant prospects for gravel working, though they might in places form the basis for small local pits.

### Alluvium and River Terraces

The alluvium and terraces of the rivers Dee and Gowy are described on pp.80–83. There are few records on the thickness of the deposits; a well aat Meadowhouse Farm [390 570] is said to have proved nearly 6 m of alluvial sands and gravels. Elsewhere gravel does not figure largely in descriptions of either river terraces or floodplain alluvium, and there is commonly much material of clay grade, together with sand and silt.

The alluvium and terraces of the rivers Weaver and Dane contain a few lenses of shingle, but these deposits are in general too thin to offer serious economic prospects.    BJT

## BRICK CLAYS

The only deposit that has been extensively exploited as a source of brick clay is boulder clay, but the industry has been extinct for some 25 years or more. Pits that seem to have been active to within a few years of the commencement of the primary six-inch survey in 1953 include ones at Hoole Bank Brickworks [433 692], north-east of Chester (see p. 69); at the brickworks [445 660] east of Chester on the south side of the A51 at Littleton (see p. 69); at Tattenhall Road Brickworks [495 607] NNE of Tattenhall (see p. 70); and at Clayton Brickworks [582 589] near Alpraham (see p. 71). At the time of Strahan's survey over a century ago the industry was very active in the vicinity of Chester and some of his observations on the various sources then being exploited are reproduced on p. 69. In the same context (Strahan, 1882, p. 36), he mentioned that the 'Keuper' Marls have been manufactured into bricks at Cuddington Waste.

## BUILDING STONE

The red and grey 'freestones' of the Helsby Sandstone are still much in evidence in older domestic buildings, churches, bridges and retaining walls throughout west Cheshire, and there are surviving old buildings in Chester itself and the surrounding villages that were constructed from the local Pebble Beds. Flaggy sandstones formerly used as hearth-stones, doorsteps and window-sills were often obtained from quarries in the Tarporley Siltstones. One of the last major civil engineering structures to be built of red

'freestone' is the Liverpool Corporation Water Tower at Norton [552 815], just north of this district. It is a remarkable example of a circular tower in which exceptionally massive rough-hewn blocks of Helsby Sandstone make a huge plinth that supports the upper structure.

The sandstone quarrying industry, still very active in the latter part of the nineteenth century, died out in the early part of this century, and the following extracts from Strahan (1882, pp. 3–9) are probably worth recording here.

### Pebble Beds

'The Pebble Beds have been quarried in many places for building purposes, but their inferiority for this purpose is shown by the condition of Chester Cathedral (before restoration), and St John's Church tower. A great proportion of the stone used in the Cathedral and the old Abbey buildings adjoining it was obtained from quarries between the Northgate Street and Windmill Lane, and in places within the walls on the east side of the Northgate Street'.

'It is probable that many of the quarries in Chester are of great antiquity as this stone was used by the Romans in portions of their buildings; but that they were aware of its inferiority as a building material is shown by their having used large and well-trimmed blocks of the sandstones of the Keuper Basement Beds for the facing of their city wall. In mediaeval times the local stone was more universally worked, and little more than crumbling ruins remain of the buildings constructed of this poor material. Most of the villages of the Bunter area are situated on the hills formed by the Pebble Beds, and in all of these the church and other principle (sic) buildings have been constructed of the Pebble Bed Sandstone obtained from quarries in the immediate neighbourhood. The principal quarries are in the villages of Eccleston, Christleton, Waverton, Saighton, Handley, Tattenhall, and Tarvin. The stone seems to last better when not exposed to the vitiated atmosphere of a town'.

### Helsby Sandstone

'The building-stone obtained from these beds is superior to that of the Pebble Beds of the Bunter, and has been used to replace it in the restoration of Chester Cathedral. Its usual colour is a subdued brownish red, occasionally mottled with white; more rarely it is pure white. The chief quarries are at Helsby, Manley, Simmond's Hill, Delamere, Kelsall and Peckforton'. At Helsby Quarry [490 749], 'the section in the two lower courses of sandstone in the quarry....... is as follows: Sandstone, good building stone 25 + ft; (on) soft, white, current-bedded sand with two 8-inch bands of dark red shale 5 ft; ditto with irregular red seams 10 ft; dark red shale with bullions 7 ft; (over) sandstone, good building stone 40 + ft'.

At Manley Quarry [498 720]: 'The Manley Stone occurs in a bed about 45 ft thick, the best quality being at the bottom. The grain is coarse and sharp, and generally too loose to make a first-class building material. Rolled lumps of green shale and small pebbles of quartzite occur in it'. ...... 'The stone has been used in the Grosvenor Bridge and Chester Castle, including the monoliths supporting the portico, and in parts of Eaton Hall'.

At Delamere the quarry noted by Strahan is almost certainly that now called Eddisbury Quarry [557 693]. Here the stone, 'is traversed by strong joints running S.25°E..... the following is the descending section: Worthless stone, full of pebbles and soft fragments 0 to 4ft; brown sandstone, with a few small quartzite pebbles 16 ft; coarse loose grit with quartzite pebbles and very numerous cavities..... 0 to 4 ft; brown sandstone 18 + ft'.

The Peckforton area is mentioned by Poole and Whiteman (1966, p.97) as follows: 'The Keuper Sandstone Conglomerate also provides a good durable building stone and has been widely used for farm buildings and stone walls. There are numerous old quarries in this formation in the Peckforton Hills and Peckforton Castle in the adjacent Chester (109) district is built of it'.

### Tarporley Siltstones

Brief references to the use as building stone of material from near the base of the 'Waterstones' are given by Strahan. Commenting on flaggy sandstones in a section, near Kelsall, defined as 'the quarry on Longley Hill' he states (p.11), 'The stone is of good quality; it splits readily into slabs, and is worked into hearth-stones, window sills etc. The surfaces of the shales show beautifully preserved ripple-marks'. Similarly with reference (p.12) to other quarries defined as 'on Luddington Hill'; 'Near Utkinton'; and 'near Tarporley on the Chester Road'; he states 'The flags and sandstones are worked up into window-sills'.

## CIVIL ENGINEERING ASPECTS

### Salt subsidence

The details of subsidence due to salt solution in this district are given on pp. 41–56. The engineering aspect chiefly concerns foundation stability, and the precautions which must be taken against damage to buildings, bridges and other structures.

The possibility of salt solution, and therefore of subsidence, exists throughout the salt subcrop or 'wet' rockhead area, but the subsidence risk is not the same in all parts of the subcrop. As is pointed out on p. 41, the routes of brine movement, and the points of ingress of fresh replacement water, tend to become firmly established, and unless there is some fundamental change in the hydraulic regime in the ground, such as might be caused by the establishment of a new brine well or by a large increase in existing abstractions, the chances against the development of a completely new area of subsidence are quite high.

The Cheshire Brine Subsidence Compensation Board has the statutory duties of examining claims for damage due to salt solution subsidence in the area of the saltfields, and of paying compensation in approved cases, the money being secured by a levy on the brine producers. The sites of claims registered and paid by the Board form the basis of a map showing the areas of greatest recorded subsidence risk. Within these 'yellow areas', so-called because of the colour used on the key map, the Board may require that the design of new structures should be approved by them if the owners

are to be covered by the compensation scheme. Requirements may include rafted or jackable foundations or other structural strengthening. Since the 'yellow areas' are based on the numbers of actual claims, there may be subsidences known to the Board, particularly away from built-up areas, which are not included in the 'yellow areas'.

Framed and rafted structures have been in use for many years in those parts of Winsford and Northwich most at risk from subsidence. As the ground fell away, jacks beneath the buildings raised them to their former level (Plate 5), while the surrounding ground was made up with fill. Similar lifting facilities have been incorporated in bridge footings. The dramatic subsidences affecting ground and property in Northwich, described and illustrated by Ward (1900) and Calvert (1915), are unlikely to be repeated. The circumstances, which at one time involved the deliberate flooding and subsequent pumping of brine from old rock-salt mines, would not be allowed today. At the same time the accidental flooding of an existing salt mine would put the ground above it at serious risk.

A secondary consequence of past salt-solution is that the collapsed strata and solution residue may have a lower bearing strength than undisturbed Mercia Mudstones. This effect exists to a greater or lesser degree over the whole of the salt subcrop and for some distance beyond the up-dip edge (A–C on Figure 15). E. H. Sale of Imperial Chemical Industries, Ltd., (1962) reported, following a gravimetric survey of part of the Winsford area, that when mudstone is shattered by the removal of underlying salt its effective density is reduced. Since this effect is presumably due to the introduction of voids which are not entirely removed by subsequent re-packing, the compressive strength of the mudstone is probably affected also.

## HYDROGEOLOGY

The annual rainfall of the district varies from less than 70 cm in the low lying areas of the Dee and Weaver catchments to 85 cm in the Delamere and Peckforton Hills. The area to the east of the Mid Cheshire Ridge is principally drained by the Weaver. West of the Ridge the low lying ground is drained by the Gowy and the Dee. The Dee and its tributaries (Hydrometric Area 67) are administered by the Welsh Water Authority whilst the catchments of the Weaver and Gowy (Hydrometric Area 68) are under the control of North West Water Authority.

Public water supplies to Chester (Chester Water Co.) are provided by abstractions from the Dee and from a borehole at Plemstall [450 700]. Regulation of flows in the upper catchment of the River Dee makes large quantities of water available to the North West Water intake works at Huntington. Supplies from the Dee, together with borehole sources in the Gowy catchment, meet local demands and provide large quantities of water for the Merseyside conurbation to the north of the district. Boreholes in the Thornton-le-Moors area meet the needs of the refinery complex at Stanlow on the south bank of the Mersey.

The major aquifer is the Sherwood Sandstone Group, which is tapped by a number of major producing boreholes that supply substantial parts of mid-Cheshire.

## Carboniferous

Local farm and estate supplies are provided by shallow wells and boreholes penetrating the Erbistock Formation and overlying superficial deposits of the Milton Green inlier. Yields are small and the waters are mineralised. A spring supply [457 565], west of Aldersey Hall, yields the following analysis:

|      | Meg/l   |
|------|---------|
| Ca   | 23.70   |
| Mg   | 18.92   |
| Na   | 243.60  |
| K    | 0.75    |
| $SO_4$ | 11.14 |
| Cl   | 282.10  |
| $NO_3$ | 0.04  |
| Alk  | 3.00    |
| pH.  | 7.3     |

Waters of this type are present throughout the Erbistock succession.

## Permian and Triassic

Groundwater resources in the Permian and Triassic rocks are largely restricted to the Sherwood Sandstone Group which crops out over the western half of the district. Groundwaters in the Mercia Mudstone Group are very limited in quantity, and where present are generally highly mineralised.

Boreholes in the Sherwood Sandstone are licenced for more than 58 000 $m^3$/d for public supply, and for industrial and agricultural use. Major sources are listed below:

| Site              | Nat. Grid reference | User             |
|-------------------|---------------------|------------------|
| Plemstall         | 450 700             | Chester Water Co.|
| Newton Hollows    | 526 743             | N.W.W.A.         |
| Foxhill           | 506 748             | N.W.W.A.         |
| Ashton            | 505 689             | N.W.W.A.         |
| Mouldsworth       | 503 704             | N.W.W.A.         |
| Manley Common     | 515 717             | N.W.W.A.         |
| Manley Quarry     | 490 719             | N.W.W.A.         |
| Organsdale        | 553 683             | N.W.W.A.         |
| Sandyford         | 568 672             | N.W.W.A.         |
| Delamere          | 561 677             | N.W.W.A.         |
| Eddisbury         | 557 694             | N.W.W.A.         |
| Eaton             | 567 634             | N.W.W.A.         |
| Cotebrook         | 572 657             | N.W.W.A.         |
| Tattenhall        | 521 583             | S.T.W.A.         |
| Peckforton        | 528 556             | S.T.W.A.         |
| Thornton-le-Moors | 431 746             | Industrial       |

In addition to the above pumping stations there are numerous small sources providing supplies for agricultural and domestic use.

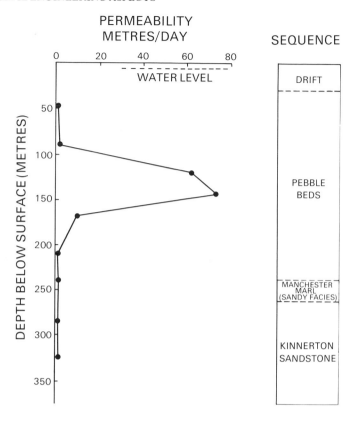

**Figure 25**   Packer tests, Priors Heyes Borehole

The major boreholes variously penetrate all the units of the Sherwood Sandstone Group including the Helsby Sandstone, the Upper Mottled Sandstone and the Pebble Beds. Individual yields vary between 2000 and 10 000 $m^3$/d. Pumping tests indicate that transmissivity values range between 100 and 500 $m^2$/d, with exceptional high values of up to 12 000 $m^2$/d in the Delamere area. Laboratory determinations of the intergranular permeability of core samples (Campbell, 1982) have given values that are rarely high enough to explain these yields. Packer testing (Brassington and Walthall, 1985) has been used to establish the vertical distribution of permeability in water-level monitoring boreholes at Priors Heyes [513 664], Burton [506 646] and Organsdale [551 683]. The results of these tests at Priors Heyes are shown in Figure 25. The high permeability zones correspond to fissuring recorded by closed-circuit television inspection. Fissuring has been shown to be common to depths of about 150 m below surface, and makes a major contribution (up to 80 per cent) to the yield of the boreholes.

The distribution of peizometric head established by recent monitoring of the aquifer (North West Water Authority, *in press*) is illustrated in Figure 26. Contours of the peizometric surface indicate movement across the Permian–Carboniferous boundary of the Milton Green inlier, and suggest a close relationship between ground and surface waters in the middle reaches of the River Gowy.

Chemical analyses of 96 samples from sources in or near this outcrop of the Sherwood Sandstone Group give the following values for 'fresh' waters in the aquifer system:

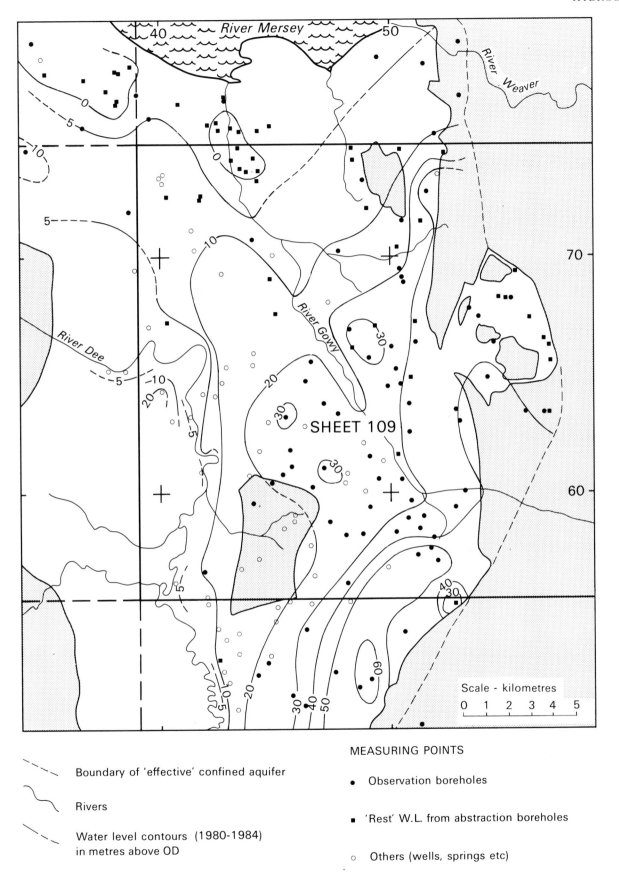

**MEASURING POINTS**

--- --- Boundary of 'effective' confined aquifer

⁀⁀⁀ Rivers

-⁀-⁀ Water level contours (1980-1984)
in metres above OD

● Observation boreholes

■ 'Rest' W.L. from abstraction boreholes

○ Others (wells, springs etc)

**Figure 26**   Sherwood Sandstone aquifer: west Cheshire—water level contours (1980–1984)

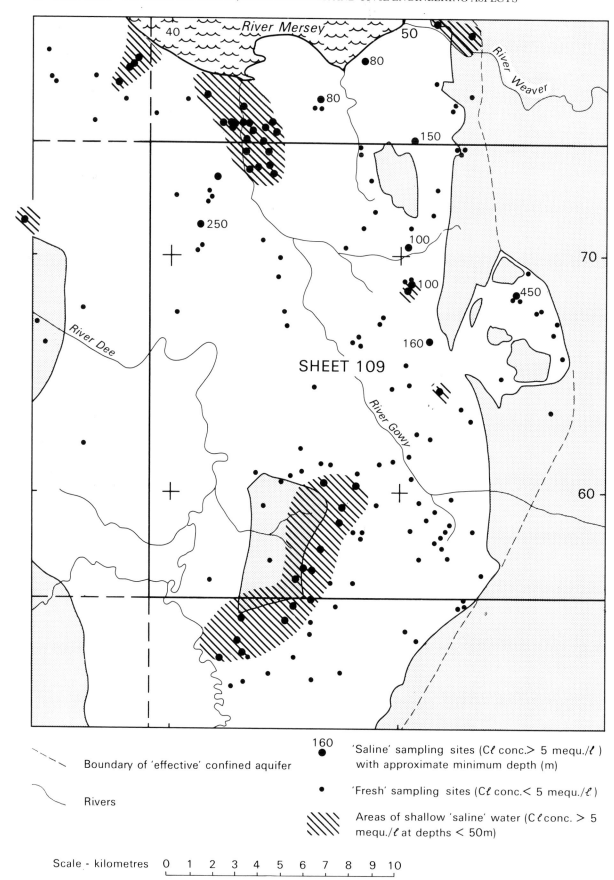

**Figure 27**  Sherwood Sandstone aquifer: west Cheshire — distribution of chloride-rich waters

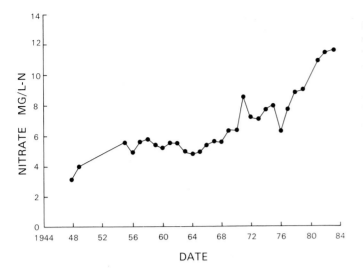

**Figure 28** Nitrate variations, Sandyford Borehole

|     | range (meg/l) | mean (meg/l) |
| --- | --- | --- |
| Ca | 1.76 – 3.67 | 2.64 |
| Mg | 0.40 – 1.87 | 0.98 |
| Na | 0.43 – 0.69 | 0.55 |
| K | 0.04 – 0.06 | 0.045 |
| $SO_4$ | 0.17 – 0.88 | 0.53 |
| Cl | 0.51 – 0.92 | 0.74 |
| $NO_3$ | 0.26 – 0.72 | 0.44 |
| $SiO_2$ | 7.45 – 12.83 | 10.36 |
| Alk | 2.11 – 3.26 | 2.56 |
| pH. | 7.21 – 7.72 | 7.64 |

Saline waters are found in the vicinity of the Milton Green inlier and near Thornton-le-Moors (Figure 27). In addition, saline waters have been found below 'fresh' groundwater at depths shown on Figure 27. Analyses indicate that the saline waters in the Thornton-le-Moors borehole originate from the ingress of modern Mersey estuary waters induced by pumping (North West Water Authority, *in press*). Chemistry of the deep and inland saline waters suggests that they were formed by the solution of halite, followed by dilution with meteoric water. (North West Water Authority, *in press*): these waters lie below the zone of active groundwater movement in areas where groundwater movement is limited by geological conditions.

Increasing use of nitrogenous fertilizers has led to a build-up of nitrates in groundwater, since nitrates are carried down to the water table by natural recharge. Figure 28 shows the long-term increase in nitrate concentration of the waters pumped from the Sandyford Borehole.

## Superficial deposits

Numerous shallow wells and boreholes penetrate the superficial deposits, especially the sands and gravels, and provide small supplies for agricultural and domestic use. In the past, public supplies were taken from springs and shallow wells in these deposits, but their variable bacteriological and chemical quality has led to these sources being abandoned.

The extensive sand and gravel sheet to the east of the Delamere Hills is in hydraulic continuity with the Sherwood Sandstone west of the East Delamere Fault. Regional groundwater flow modelling (North West Water Authority, *in press*) suggests that the sands and gravels contribute some 5000 m³/d to the resources of the Sherwood Sandstone aquifer.

**Plate 5**   The Wheatsheaf Hotel in Castle St., Northwich raised on jacks in 1960 to compensate for salt subsidence

# REFERENCES

ARTHURTON, R. S. 1980. Rhythmic sedimentary sequences in the Triassic Keuper Marl (Mercia Mudstone Group) of Cheshire, northwest England. *Geol. J.*, Vol. 15, 43–58.

AUDLEY-CHARLES, M. G. 1970. Triassic palaeogeography of the British Isles. *Q. J. Geol. Soc. London*, Vol. 126, 48–89.

BAILEY, E. B. 1926. Subterranean penetration by a desert climate. *Geol. Mag.*, Vol. 63, 276–280.

BINNEY, E. W. 1841. Sketch of the geology of Manchester and its vicinity. *Trans. Manchester Geol. Soc.*, Vol. 1, 35–66.

— 1862. The geology of Manchester and its neighbourhood. *Trans. Manchester Geol. Soc.*, Vol. 3, 350–65.

BOULTON, G. S. 1970a. On the origin and transport of endglacial debris in Svalbard glaciers. *J. Glaciol.*, Vol. 9, 213–229.

— 1970b. On the deposition of subglacial and melt-out tills at the margins of certain Svalbard glaciers. *J. Glaciol.*, Vol. 9, 231–245.

— 1972. Modern Arctic glaciers as depositional models for former ice-sheets. *J. Geol. Soc. London*, Vol. 128, 361–393.

— and WORSLEY, P. 1965. Late Weichselian glaciation of the Cheshire-Shropshire basin. *Nature, London*, Vol. 207, 704–706.

BRASSINGTON, F. C. and WALTHALL, S. 1985. Field techniques using borehole packers in hydrogeological investigations. *Q.J. Eng. Geol.*, Vol. 18, 181–193

CALVER, M. A. 1968. Distribution of Westphalian marine faunas in northern England and adjoining areas. *Proc. Yorkshire Geol. Soc.*, Vol. 37, 1–72.

— and SMITH, E. G. 1974. The Westphalian of North Wales. Pp. 169–183 in *The Upper Palaeozoic and Post-Palaeozoic rocks of Wales*. OWEN, T. R. (editor). (Cardiff: University of Wales.)

CALVERT, A. F. 1915. *Salt in Cheshire.* xxiii + 1206 pp. (London: E and F.N. Spon.)

CAMPBELL, J. E. 1982. Permeability characteristics of the Permo Triassic sandstones of the Lower Mersey Basins. Unpublished PhD thesis, University of Birmingham.

COLTER, V. S. and BARR, K. W. 1975. Recent developments in the geology of the Irish Sea and Cheshire basins. Pp. 61–75 in *Petroleum and the Continental Shelf of north-west Europe, Vol. 1 Geology*. WOODLAND, A. W. (editor). ix + 501 pp. (London: Applied Science Publishers.)

COOPE, G. R. 1959. A late Pleistocene insect fauna from Chelford, Cheshire. *Proc. Roy. Soc. London*, Ser. B, Vol. 151, 70–86.

— MORGAN, A. and OSBORNE, P. J. 1971. Fossil coleoptera as indicators of climatic fluctuations during the last glaciation in Britain. *Palaeogeog., Palaeoclim., Palaeoecol.*, Vol. 10, 87–101.

DAWKINS, W. B. 1878. The mammoth at Northwich. *Trans. Manchester Geol. Soc.*, Vol. 15, 55–57.

DEBENHAM, F. 1920. A new mode of transportation by ice: the raised marine shells of South Victoria Land (Antarctica). *Q. J. Geol. Soc. London*, Vol. 75, [for 1912], 51–76.

DERANCE, C. E. 1870. On the glacial phenomena of western Lancashire and Cheshire. *Q. J. Geol. Soc. London*, Vol. 26, 641–655.

— 1876–1879. Reports of the committee for investigating the circulation of underground waters in New Red Sandstone and Permian formations of England. *Reps. Br. Assoc. for the Advancement of Science*, for 1875, p. 114; for 1876, p. 95; for 1877, p. 56; for 1878, p. 382; and for 1879, p. 115.

— 1895. Results of the Salt Union boring at Marston near Northwich. *Trans. Manchester Geol. Soc.*, Vol. 22, 269–302.

DICKINSON, J. 1873. *Report on the subject of the landslips in the Salt Districts, to H.M. Sec. of State for the Home Department.* London.

— 1882. Pp. 52–117 in *Reports of the Inspectors of Mines to Her Majesty's Secretary of State for the Year 1881*. London.

EARP, J. R. 1958. Mineral veins of the Minera-Maeshafn district of North Wales. *Bull. Geol. Surv. G.B.*, No. 14, 44–69.

EGERTON, P. G. 1835. On a bed of gravel containing marine shells of recent species at 'The Willington' in Cheshire. *Proc. Geol. Soc. London*, Vol. 2, p. 189.

— 1836. A notice on the occurrence of marine shells in a bed of gravel at Norley Bank, Cheshire. *Proc. Geol. Soc. London*, Vol. 2, p. 415.

— 1838. On two casts in sandstone of impressions of the hind foot of a gigantic *Cheirotherium* from the New Red Sandstone of Cheshire. *Proc. Geol. Soc. London*, Vol. 3, 11–15.

—and TAYLOR, J. 1839. Letters on a slab of sandstone containing impressions of *Cheirotherium Herculis* at the house of Mr Potts in Cheshire. *Proc. Geol. Soc. London*, Vol. 3, 100–101.

EVANS, W. B. 1966. P. 54 in *Annual report for 1965*. INSTITUTE OF GEOLOGICAL SCIENCES. (London: HMSO.)

— 1970. The Triassic salt deposits of north-western England. *Q. J. Geol. Soc. London*, Vol. 126, 103–123.

— WILSON, A. A., TAYLOR, B. J. and PRICE, D. 1968. Geology of the country around Macclesfield, Congleton, Crewe and Middlewich (Sheet 110). *Mem. Geol. Surv. G.B.*

FISHER, M. J. 1972a. A record of palynomorphs from the Waterstones (Triassic) of Liverpool. *Geol. J.*, Vol. 8, No. 1, 17–22.

— 1972b. The Triassic palynofloral succession in England. *Geoscience and Man*, Vol. 4, 101–109.

FITCH, F. J., MILLER, J. A. and THOMPSON, D. B. 1966. The palaeogeographic significance of isotopic age determinations on detrital micas from the Triassic of the Stockport-Macclesfield district, Cheshire, England. *Palaeogeogr. Palaeoclim. Palaeoecol.*, Vol. 2, 281–312.

GALE, I. N., EVANS, C. J., EVANS, R. B., SMITH, I. F. HOUGHTON, M. T. and BURGESS, W. G. 1984. The Permo-Triassic aquifers of the Cheshire and West Lancashire basins. *Invest. Geotherm. Potential UK. (Keyworth: British Geological Survey.)*

GEORGE, T. N., JOHNSON, G. A. L., MITCHELL, M., PRENTICE, J. E., RAMSBOTTOM, W. H. C., SEVASTOPULO, G. D. and WILSON, R. B. 1976. A correlation of Dinantian rocks in the British Isles. *Spec. Rep. Geol. Soc. London*, No. 7, 87 pp.

HICKLING, C. 1918a. The geology of Manchester as revealed by borings. *Trans. Inst. Min. Eng.*, Vol. 54, 367–417.

— 1918b. The geology of Manchester as revealed by borings. *Trans. Manchester Geol. Soc.*, Vol. 35, 189–239.

Hicks, H. 1885. On some recent searches in bone caves in Wales. *Proc. Geol. Assoc.*, Vol. 9, 1–20.

Hull, E. 1860. ' On the new subdivisions of the Triassic rocks of the central counties. *Trans. Manchester Geol. Soc.*, Vol. 2, 22–34.

— 1864. Geology of the country around Oldham. *Mem. Geol. Surv. G.B.*

— 1869. The Triassic and Permian rocks of the midland counties of England. *Mem. Geol. Surv. G.B.*

Ireland, R. J., Pollard, J. E., Steel, R. J. and Thompson, D. B. 1978. Intertidal sediments and trace fossils? from the Waterstones (Scythian-Anisian) at Daresbury, Cheshire. *Proc. Yorkshire Geol. Soc.*, Vol. 41, 339–436.

Jones, R. C. B., Tonks, L. H. and Wright, W. B. 1938. Wigan District. *Mem. Geol. Surv. G.B.*

Jones , T. A. 1935. The Mouldsworth boreholes of the West Cheshire Water Board. *Proc. Liverpool Geol. Soc.*, Vol. 16, 266–269.

Kent, P. E. 1975. The tectonic development of Great Britain and the surrounding seas. Pp. 3–29 in *Petroleum and the Continental Shelf, Vol. 1 Geology*. Woodland, A. W. (editor). ix + 501 pp. (London: Applied Science Publishers.)

Mackintosh, D. 1873. Observations on the more remarkable boulders of the north-west of England and the Welsh Borders. *Q. J. Geol. Soc. London*, Vol. 29, 351–360.

— 1880. On the correlation of the drift deposits of the north-west of England with those of the Midlands and eastern counties. *Q. J. Geol. Soc. London*, Vol. 36, 179–188.

McMillan, N. F. 1958. A Holocene deposit near Chester containing *Dreissena*. *Liverpool and Manchester Geol. J.*, Vol. 2, 82–85.

McQuillin, R. 1964. Geophysical investigation of seismic shot-holes in the Cheshire basin. *Bull. Geol. Surv. G.B.*, No. 21, 197–203.

North West Water Authority. *In press.* Saline groundwater investigation West Cheshire Permo Triassic sandstone aquifer. (Warrington: North West Water Authority.)

Notholt, A. J. G. and Highley, D. E. 1973. Salt. *Mineral Resources Consultative Committee; Mineral Dossier* No. 7. 36 pp.

Ormerod, G. W. 1848. Outline of the principal geological features of the saltfield of Cheshire and the adjoining districts. *Q. J. Geol. Soc. London*, Vol. 4, pp. 262 and 268–288.

— 1869. On the Salt Field and New Red Sandstone of Cheshire. *Trans. Manchester Geol. Soc. for 1843*, Vol. 8, 25–30.

Owen, T. R. 1976. *The geological evolution of the British Isles.* (Oxford: Pergamon Press.)

Penny, L. F. 1964. A review of the last glaciation in Great Britain. *Proc. Yorkshire Geol. Soc.*, Vol. 34, 387–411.

Pollard, J. B. 1981. A comparison between the Triassic trace fossils of Cheshire and south Germany. *Palaeontology*, Vol. 24, 555–588.

Poole, E. G. and Whiteman, A. J. 1966. The geology of the country around Nantwich and Whitchurch. *Mem. Geol. Surv. G.B.*, Sheet 122.

Pugh, W. J. 1960. Triassic salt: discoveries in the Cheshire-Shropshire basin. *Nature, London*, Vol. 187, 278–279.

Reade, T. M. 1874. The drift beds of the north-west of England: Part I, Shells of the Lancashire and Cheshire low-level boulder clay and sands. *Q. J. Geol. Soc. London*, Vol. 30, p. 27.

— 1888. An estimate of post-Glacial time. *Q. J. Geol. Soc. London*, Vol. 44, p. 291.

Rowlands, B. M. 1971. Radiocarbon evidence of the age of an Irish Sea glaciation in the Vale of Clwyd. *Nature, London, Physical Science*, Vol. 230, 9–11.

Sedgwick, A. 1835. On the geological relations and internal structure of the Magnesium Limestone and the lower portions of the New Red Sandstone Series in their range through Nottinghamshire, Derbyshire, Yorkshire and Durham to the Southern extremity of Northumberland. *Trans. Geol. Soc. London (Ser. 2)*, Vol. 3, 37–124.

Sherlock, R. L. 1921. Rock salt and brine. *Econ. Mem. Geol. Surv. G.B.*, Vol. 18, 121 pp.

— 1938. A boring at Booth's Mill, Knutsford. *Summ. Prog. Geol. Surv. G.B. for 1936*, 52–53.

Shone, W. 1874. Discovery of foraminifera, etc, in the boulder clays of Cheshire. *Q. J. Geol. Soc. London*, Vol. 30, 181–185.

— 1878a. How we found the microzoa in the boulder clay of Cheshire. *Midland Naturalist*, Vol. 1, p. 292.

— 1878b. On the glacial deposits of west Cheshire together with lists of the fauna found in the drifts of Cheshire and adjoining counties. *Q. J. Geol. Soc. London*, Vol. 34, 383–397.

Shotton, F. W., Blundell, D. J. and Williams, R. E. G. 1969. Birmingham University Radiocarbon dates III. *Radiocarbon*, Vol. 11, 263–270.

Simpson, I. M. 1959. The Pleistocene succession in the Stockport and south Manchester area. *Q. J. Geol. Soc. London*, Vol. 115, 107–121.

— and West, R. G. 1958. On the stratigraphy and palaeobotany of a Late-Pleistocene organic deposit at Chelford, Cheshire. *New Phytologist*, Vol. 57, 239–250.

Stephens, J. V. 1958. P. 34 in *Summary of Progress for 1957*. Geological Survey of Great Britain. (London: HMSO.)

— 1959. P. 35 in *Summary of Progress for 1958*. Geological Survey of Great Britain. (London: HMSO.)

Strahan, A. 1882. Geology of the neighbourhood of Chester. (Old Series Quarter Sheet 80 SW.) *Mem. Geol. Surv. G.B.*

— 1905. *Report of the British Association for the Advancement of Science for 1904* (Cambridge), Sect. C, 532–541.

Taylor, B. J. 1958. Cemented shear-planes in the Middle Sands of Lancashire and Cheshire. *Proc. Yorkshire Geol. Soc.*, Vol. 31, 359–365.

— 1961. The stratigraphy of exploratory boreholes in the West Cumberland Coalfield. *Bull. Geol. Surv. G.B.*, No. 17, 1–74.

— Price, R. H. and Trotter, F. M. 1963. Geology of the country around Stockport and Knutsford (Sheet 98). *Mem. Geol. Surv. G.B.*

Thompson, D. B. 1969. Dome-shaped aeolian dunes in the Frodsham member of the so-called 'Keuper' Sandstone Formation (Scythian – ?Anisian: Triassic) at Frodsham, Cheshire (England). *Sediment. Geol.*, Vol. 3, 263–289.

— 1970a. Sedimentation of the Triassic (Scythian) red pebbly sandstone in the Cheshire basin and its margins. *Geol. J.*, Vol. 7, 183–216.

— 1970b. The stratigraphy of the so-called Keuper Sandstone Formation (Scythian – ?Anisian) in the Permo-Triassic Cheshire basin. *Q. J. Geol. Soc. London*, Vol. 126, 151–181.

— and Worsley, P. 1966. A late Pleistocene molluscan fauna from the drifts of the Cheshire Basin. *Geol. J.*, Vol. 5, 197–207.

Trimmer, J. 1833. Discovery of marine shells of existing species on the left bank of the River Mersey, and above the level of high water mark. *Proc. Geol. Soc. London*, Vol. 1, p. 419.

— 1851. On the erratic Tertiaries bordering the Pennine Chain between Congleton and Macclesfield; and on the scratched detritus of the till. *Q. J. Geol. Soc. London*, Vol. 7, 201–207.

TROTTER, F. M. 1953. Reddened beds of Carboniferous age in north-west England and their origin. *Proc. Yorkshire Geol. Soc.*, Vol. 29, 1–20.

TRUEMAN, A. E. 1947. Stratigraphical problems in the coalfields of Great Britain. *Q. J. Geol. Soc. London*, Vol. 103, lxv–civ.

TUCKER, R. M. and TUCKER, M. E. 1981. Evidence of synsedimentary tectonic movements in the Triassic halite of Cheshire. *Nature, London*, Vol. 290, 495–496.

WARD, T. 1873. The Cheshire Salt District. *Proc. Lit. Phil. Soc. Liverpool*, Vol. 27, 39–63.

— 1898. The rock salt deposits of Cheshire and the result of their exploitation. *Trans. Manchester Geol. Soc.*, Vol. 25, 274–306 and 530–567.

— 1900. The subsidence in and around the town of Northwich in Cheshire. *Trans. Inst. Min. Eng.*, Vol. 19, 241–264.

WARRINGTON, G. 1967. Correlation of the Keuper Series of the Triassic by miospores. *Nature, London*, Vol. 214, 1323–1324.

— 1970a. The 'Keuper' Series of the British Trias in the northern Irish Sea and neighbouring areas. *Nature, London*, Vol. 226, 254–256.

— 1970b. The stratigraphy and palaeontology of the 'Keuper' Series of the central midlands of England. *Q. J. Geol. Soc. London*, Vol. 126, 183–223.

— 1974. Studies in the palynological biostratigraphy of the British Trias. I. Reference sections in west Lancashire and north Somerset. *Rev. Palaeobot. Palynol.*, Vol. 17, 133–147.

— AUDLEY-CHARLES, M. G., ELLIOTT, R. E., EVANS, W. B., IVIMEY-COOK, H. C., KENT, P. E., ROBINSON, P. L., SHOTTON, F. W. and TAYLOR, F. M. 1980. A correlation of Triassic rocks in the British Isles. *Geol. Soc. London, Spec. Rep.*, No. 13, 78 pp.

WEDD, C. B., SMITH, B., SIMMONS, W. C. and WRAY, D. A. 1923. The geology of Liverpool, with Wirral and part of the Flintshire coalfield (Sheet 96). *Mem. Geol. Surv. G.B.*

— and KING, W. B. R. 1924. The geology of the country around Flint, Hawarden and Caergwrle. (Sheet 108). *Mem. Geol. Surv. G.B.*

— SMITH, B. and WILLS, L. J. 1927. The geology of the country around Wrexham. Part 1. Lower Palaeozoic and Lower Carboniferous Rocks. *Mem. Geol. Surv. G.B.*

— 1928. The geology of the country around Wrexham. Part 2. Coal Measures and newer formations. *Mem. Geol. Surv. G.B.*

WHITE, P. H. N. 1949. Gravity data obtained in Great Britain by the Anglo-American Oil Co. Ltd. *Q. J. Geol. Soc. London*, Vol. 104, 339–364.

WILLS, L. J. 1912. Late-Glacial and post-Glacial changes in the Lower Dee valley. *Q. J. Geol. Soc. London*, Vol. 68, 180–198.

— 1956. *Concealed coalfields.* (London and Glasgow: Blackie.)

— 1970. The Triassic succession in the central Midlands. *Q. J. Geol. Soc. London*, Vol. 126, 225–283.

WOOD, A. 1937. The non-marine Lamellibranchs of the North Wales Coalfield. *Q. J. Geol. Soc. London*, Vol. 93, 1–22.

WRAY, A. D. and EARP, J. R. 1944. Water supply from underground sources of North Wales and parts of Lancashire, Cheshire and Shropshire. *Geol. Surv. Wartime Pamphlet*, No. 26, Pt. 3.

WRIGHT, W. B. 1914. *The Quaternary Ice Age.* (1st edition.) 464 pp. (London: Macmillan.)

— 1936. P. 68 in *Summary of progress for 1935*. GEOLOGICAL SURVEY OF GREAT BRITAIN. (London: HMSO.)

— 1937. Pp. 55–56 in *Summary of progress for 1936*. GEOLOGICAL SURVEY OF GREAT BRITAIN. (London: HMSO.)

— 1938. Pp. 43–44 in *Summary of progress for 1937*. GEOLOGICAL SURVEY OF GREAT BRITAIN. (London: HMSO.)

YATES, E. M. and MOSELEY, F. 1967. A contribution to the glacial geomorphology of the Cheshire Plain. *Trans. Inst. Br. Geogr.*, Vol. 42, 107–125.

# APPENDIX 1

## Generalised records of selected boreholes

Records of the more important boreholes drilled for ground-water, brine, rock salt or geological exploration are given below. The less important descriptive detail has been generalised. Abbreviated records of the older wells, boreholes and shaft sinkings in the district up to 1943 were published in Wray and Earp (1944, pp. 14–24). Boreholes drilled since the 1939–45 war are published here in fairly detailed form for the first time. In a few cases the classification of the strata in this appendix differs slightly from that adopted by Wray and Earp, and this reflects changes in our knowledge of the geology of the district since that time. In each case the serial number of the borehole in the British Geological Survey record system is given in brackets after the name. National Grid references are all in the 100-kilometre square SJ.

### Ashton No. 3 Observation Borehole (56 NW/13)
Surface level about 39 m above OD
National Grid reference 5051 6902
Drilled 1969 for Mid Cheshire Water Board
Cores examined by A. A. Wilson

| | Thickness m | Depth m |
|---|---|---|
| QUATERNARY | | |
| Soil and fine to-coarse-grained sand | 41.00 | 41.00 |
| Sand and gravel | 3.00 | 44.00 |
| UPPER MOTTLED SANDSTONE | | |
| Sandstone, fine-grained, purplish brown | 10.50 | 54.50 |
| Sandstone, fine- to medium-grained, reddish brown, a few quartzite pebbles | 5.00 | 59.50 |
| Sandstone, fine- to medium-grained, reddish brown, micaceous | 3.00 | 62.50 |
| Sandstone, very fine-grained, micaceous, reddish brown, ripple marked at 63 m | 4.50 | 67.00 |
| Sandstone, fine- to medium-grained　　*no core* | 10.75 | 77.75 |
| Sandstone, fine- to medium-grained, becoming coarseer and more micaceous downwards | 4.25 | 82.00 |
| Sandstone, fine- to medium-grained　　*no core* | 10.00 | 92.00 |
| Sandstone, fine-grained, reddish brown, with 'millet-seed' grains, a grey and purple micaceous bed 51 cm at 92.5 m, scattered quartz pebbles below this depth | 3.00 | 95.00 |
| Mudstone, red, micaceous and grey-green siltstone 15 cm on fine- to medium-grained, reddish brown sandstone | 3.50 | 97.50 |
| Sandstone, fine-grained　　*no core* | 3.00 | 100.50 |
| Sandstone, fine-grained, reddish brown, becoming micaceous downwards | 3.00 | 103.50 |
| Sandstone, fine-grained　　*no core* | 10.50 | 114.00 |
| Sandstone, fine- to very fine-grained, reddish brown | 3.25 | 117.25 |
| Sandstone, fine-grained, reddish brown, 'millet-seed' grains towards base | 7.75 | 125.00 |
| Sandstone fine- to medium-grained, reddish brown, some 'millet-seed' grains, becoming abundant below 129.5 | 12.00 | 137.00 |
| Sandstone, fine- to medium-grained, reddish brown, with some 'millet-seed' grains | 15.25 | 152.25 |

### Austin's Springs Borehole (56 NE/11)
Surface level 63.31 m above OD
National Grid reference 5952 6722
Drilled 1899 for Winsford Urban District Council

| | Thickness m | Depth m |
|---|---|---|
| QUATERNARY | | |
| Soil | 0.46 | 0.46 |
| Sand, grey and red | 7.77 | 8.23 |
| Clay, fine plastic | 7.62 | 15.85 |
| Sand, red | 6.71 | 22.56 |
| Clay, fine plastic | 7.92 | 30.48 |
| Sand, red | 5.49 | 35.97 |
| Gravel, coarse loamy | 4.26 | 40.23 |
| Gravel, fine sandy | 6.10 | 46.33 |
| Clay, red plastic | 7.01 | 53.34 |
| Sand, red | 7.92 | 61.26 |
| Clay, sandy | 3.36 | 64.62 |
| Gravel | 0.91 | 65.53 |
| Clay, sandy | 2.75 | 68.28 |
| Gravel, sandy | 2.74 | 71.02 |
| Clay, sandy | 2.13 | 73.15 |
| Sand, red | 6.44 | 75.59 |
| Clay, plastic | 17.07 | 92.66 |
| MERCIA MUDSTONE GROUP | | |
| Marl and sandy marl, red | 75.59 | 168.25 |
| Marlstone, hard grey | 0.60 | 168.85 |
| Marl, sandy red | 24.70 | 193.55 |
| Vein of blue shale | 0.15 | 193.70 |
| Marl, sandy red | 13.62 | 207.32 |
| Marl, red; with gypsum | 4.57 | 211.89 |
| Marl, sandy red | 15.55 | 227.44 |
| Marl, grey | 4.88 | 232.32 |
| Marl, red | 2.13 | 234.45 |
| Marl, sandy strong, with thin gypsum streaks | 195.99 | 430.44 |
| ?TARPORLEY SILTSTONES | | |
| Sandstone, red | 3.04 | 443.48 |
| Marl, strong sandy | 12.04 | 455.52 |

General note: 'marl' and 'marlstone' denote types of mudstone and siltstone.

### Bowyer's Waste Borehole (56 NE/12)
Surface level 77.27 above OD
National Grid reference 5686 6882
Drilled 1958–59 for Geological Survey of Great Britain
Logged by B. J. Taylor, W. B. Evans and J. R. Earp (*summary*)

| | Thickness m | Depth m |
|---|---|---|
| QUATERNARY | | |
| Sand, coarse; stones up to 12 mm | 15.54 | 15.54 |
| Sand, coarse; very few small stones | 8.23 | 23.77 |
| Sand, coarse | 4.58 | 28.35 |
| Clay, silty reddish brown with stones | 4.57 | 32.92 |
| *Core drilling* | | |
| Boulder clay, brown; abundant erratic boulders | 0.76 | 33.68 |

|  | Thickness m | Depth m |
|---|---|---|
| **MERCIA MUDSTONE GROUP** | | |
| Laminated facies of red and grey mudstone and silty mudstone, alternating with blocky facies of red-brown silty mudstone with greenish spots and mottling; desiccation cracks and ripple marks in the laminated facies. Gypsum veinlets throughout | 22.25 | 55.93 |
| Rhythmic alternations as above, except that the laminated facies is predominantly silty, with many thin siltstone layers. Many gypsum veinlets; faults (?small) between 71.32 and 72.54 m | 45.57 | 101.50 |
| **NORTHWICH HALITE** | | |
| Breccia: angular fragments of red and grey mudstone and sandy mudstone in pulverised mudstone matrix (solution residue) | 7.77 | 109.27 |
| Mudstone and sandy mudstone, with some siltstone, both banded and blocky facies; layers of mudstone breccia with bases at 112 m, 118 m, 123 m, 126 m; halite traces at 110 m and 125 m amd a thin bed at 128 m. Dips in the bedded strata variable, with minor faulting and autobrecciation | 22.56 | 131.83 |
| Breccia of mudstone; halite inclusions near base | 5.48 | 137.31 |
| Mudstone, red brown intergrown with halite in equal proportions | 1.98 | 139.29 |
| Rock salt; mudstone inclusions near top and in lower half; gypsum traces | 13.49 | 152.78 |
| Mudstone, red blocky facies at top and base, elsewhere mudstone and silty mudstone interbanded; salt veins and inclusions | 8.94 | 161.72 |
| Rock salt; mudstone inclusions | 2.31 | 164.03 |
| Mudstone, red with large halite crystals | 0.87 | 164.90 |
| Rock salt; mudstone inclusions; gypsum traces | 11.73 | 176.63 |
| Fault gouge: pulverised red mudstone | 0.31 | 176.94 |
| Mudstone, red and grey, salt inclusions and veins; some faulting | 16.18 | 193.12 |

**Clotton (Lower House Farm) Borehole** (56 SW/8)
Surface level 48 m above OD
National Grid reference 5322 6366
Drilled 1958 for Geological Survey of Great Britain
Cores examined by B. J. Taylor

|  | Thickness m | Depth m |
|---|---|---|
| **QUATERNARY** | | |
| Grey and brown silty and sandy loam with small stones | 2.50 | 2.50 |
| Sand, buff and brown with small stones | 5.50 | 8.00 |
| **UPPER MOTTLED SANDSTONE** | | |
| Sandstone, soft, red with buff patches, medium sub-angular grains, lower part cross-bedded with a few mudstone pellets | 12.00 | 20.00 |
| Sandstone, soft, red with buff patches, medium-grained, some larger rounded grains | 2.00 | 22.00 |
| Sandstone, red with buff patches, fine to medium angular grained, mica flakes, a few red mudstone pellets | 7.00 | 29.00 |

|  | Thickness m | Depth m |
|---|---|---|
| Sandstone, soft, red, medium to coarse-grained some mica flakes | 1.00 | 30.00 |
| Alternations of coarse and fine, red and grey sandstone | 1.00 | 31.00 |
| Sandstone, soft, red, medium-grained, an 18 cm fine grey band at 32 m, a 30 cm grey-green sandstone on 3 cm of grey-green mudstone at base | 3.00 | 34.00 |
| Sandstone, soft, red, medium to coarse-grained, some rounded grains, grey micaceous bands: 28 cm at 41 m, 13 cm at 46 m and 30 cm at 48 m, bottom 30 cm harder and mottled | 17.00 | 51.00 |
| **PEBBLE BEDS** | | |
| Sandstone, hard, red with grey mottlings, an odd pebble at 53.5 m | 3.50 | 54.50 |
| Sandstone, red with harder grey beds at 56 m and at base, red mudstone pellets, an odd pebble at 56.5 m | 4.00 | 58.50 |
| Sandstone, red, medium-grained with harder grey beds at about 61 m, 63 m and at base, pebbles at 61.5 to 62 m | 7.50 | 66.00 |
| Sandstone, red, mainly hard with some softer layers, a few grey bands and patches, mudstone pellets in lowest 2 m | 7.00 | 73.00 |
| Sandstone, medium to coarse, red with finer micaceous grey beds at 78.5 m, 80 m, 83.5 m and 84.5 m, an odd pebble at 75.5 m, mudstone pellets at 83 m and 87.5 to 88 m, calcareous nodules at 84 m and 87 m | 15.50 | 88.50 |
| Sandstone, red with grey bands and patches, cross-bedded, medium and angular-grained, a few micaceous partings, pebbles in lower part and bottom metre or so coarse | 6.50 | 95.00 |
| Sandstone, grey-white, medium- to fine-grained, massive | 1.00 | 96.00 |
| Sandstone, red with grey bands and patches, medium- to coarse-grained, cross-bedded, calcareous nodules and sporadic pebbles almost throughout | 18.00 | 114.00 |

**Cotebrook Waterworks No. 2 Borehole** (56 NE/8)
Surface level 70 m above OD
National Grid reference 5715 6575
Drilled 1951 for Mid and South-east Cheshire Water Board
Cores examined by D. Thomas

|  | Thickness m | Depth m |
|---|---|---|
| **QUATERNARY** | | |
| Soil, sand and sandy marl | 3.50 | 3.50 |
| **HELSBY SANDSTONE** | | |
| Sandstone, red with grey mottling, medium- to fine-grained, many layers of brown mudstone nodules | 9.50 | 13.00 |
| Sandstone, red and grey, medium to fine | 5.00 | 18.00 |
| Sandstone, dark reddish brown, slightly shaly below 24 m | 13.00 | 31.00 |
| Sandstone, dark reddish brown with grey shaly beds | 2.50 | 33.50 |
| Sandstone, red with grey shaly bands | 27.00 | 60.50 |
| Shale, dark red, micaceous, sandy | 1.00 | 61.50 |

| | Thickness m | Depth m |
|---|---|---|
| Sandstone, red, medium-grained, shale wedges at about 63 m and a shale-fragment breccia at base | 6.00 | 67.50 |
| Sandstone, red, medium to fine, shale bands at top and at 69.5 m, a grey-green band with mudstone pellets at 69 m | 3.50 | 71.00 |
| Sandstone, brown, fine at top, coarsening downwards, pink bands with pebbles and mudstone pellets, a fine conglomerate at base | 6.00 | 77.00 |
| Sandstone, red-brown, medium-grained | 4.00 | 81.00 |
| Conglomeratic sandstone, pink with quartzite fragments and pebbles | 1.50 | 82.50 |
| Sandstone, red, medium-grained, thin coarser bands and small mudstone pellets | 8.50 | 91.00 |
| Shale, red and green, micaceous, 76 cm of sandstone with mudstone pellets in upper part | 2.00 | 93.00 |
| Sandstone, medium-grained, a thin conglomeratic band on 30 cm of brown shale at base | 1.00 | 94.00 |
| Sandstone, reddish brown, well bedded, occasional small mudstone nodules | 14.00 | 108.00 |

UPPER MOTTLED SANDSTONE

| | | |
|---|---|---|
| Sandstone, soft, red, with grey bands | 19.50 | 127.50 |
| Sandstone, brown and grey | 10.50 | 138.00 |
| Sandstone mainly soft, red, with harder grey bands | 23.00 | 161.00 |
| Sandstone, soft, red | 83.00 | 244.00 |

## Cotton Hall Borehole (46 NE/10)
Surface level 17.5 m above OD
National Grid reference 4655 6616
Drilled 1958 for Geological Survey of Great Britain
Cores examined by W. B. Evans and J. R. Earp

| | Thickness m | Depth m |
|---|---|---|
| QUATERNARY | | |
| Sand, clean | 19.1 | 19.1 |
| 'Marl', red | 1.6 | 20.7 |
| Sand, thin leached bands at about 24 m | 4.0 | 24.7 |
| PEBBLE BEDS | | |
| Sandstone, fine-grained, alternating white and red, some cross-bedding and 'millet-seed' layers, grey micaceous films at about 27.5 m | 3.8 | 28.5 |
| Sandstone, white, fine to coarse with green marl pellets, a quartz pebble at 31.7 m | 3.3 | 31.8 |
| Sandstone, top 60 cm reddish, remainder pale grey, yellow mottled, pebbles at about 32 m, 33.2 m and 35.8 m | 5.8 | 37.6 |
| Mudstone, grey, silty, micaceous | 0.5 | 38.1 |
| Sandstone, fine- to medium-grained, red with green mudstone bands at 39.5 to 40.5 m and a grey-green bed 1.1 m at 41.9 m, rare marl pellets, a quartz pebble at 44.5 m | 8.1 | 46.2 |
| Sandstone, pale grey with micaceous films, a quartz pebble at 46.3 m | 1.5 | 47.7 |
| Sandstone, purple-red, steep cross-bedding, 'marl' pellets at 49.4 m, pebbles scattered but common in bottom 60 cm | 7.2 | 54.9 |

| | Thickness m | Depth m |
|---|---|---|
| Sandstone, fine to medium-grained, much cross-bedding, purple-red with grey-green bands, a grey mudstone parting at 70.7 m, 'marl' pellets commonest at about 100 to 101 m, occasional pebble, base coarse | 48.7 | 103.6 |
| Sandstone, fine, passing down to coarse-grained, grey, micaceous with brownish bands | 0.9 | 104.5 |
| Sandstone, medium-grained reddish brown, highly cross-bedded | 0.6 | 105.1 |
| Sandstone, medium-grained grey and brown, with shale pellets, almost a shale-pellet conglomerate at 105.3 m | 1.0 | 106.1 |

## Crabtree Green Borehole (57 SE/1)
Surface level 55.78 m above OD
National Grid reference 5797 7084
Drilled 1934 for Northwich Rural District Council

| | Thickness m | Depth m |
|---|---|---|
| QUATERNARY | | |
| Sand, light red | 23.16 | 23.16 |
| Sand and gravel | 5.82 | 28.98 |
| MERCIA MUDSTONE GROUP (Lower mudstone division) | | |
| Hard red and grey shaly marl with bands of hard sandstone and gypsum; band of soft red marl 0.91 m at 114.60 m | 5.49 | 119.18 |
| Hard shaly marl with bands of soft sandstone and gypsum | 19.81 | 138.99 |
| Soft red marl | 3.05 | 142.04 |
| Alternations of hard shaly marl and soft red marl | 81.53 | 223.57 |
| Alternations of hard red shaly marl and shaly green marl; occasional thin sandstone bands; some gypsum veins | 50.88 | 274.45 |
| TARPORLEY SILTSTONES | | |
| Sandstone, grey, brown and red, alternating with red and grey sandy marl | 76.68 | 351.13 |

## Crossley Sanatorium Borehole (57 SW/14)
Surface level 128 m above OD
National Grid reference 5303 7341
Drilled 1901 for Manchester Corporation

| | Thickness m | Depth m |
|---|---|---|
| TARPORLEY SILTSTONES | | |
| Soil and grey marlstone | 2.1 | 2.1 |
| Marl, red | 7.0 | 9.1 |
| Rock, red and grey | 4.0 | 13.1 |
| Marl, red with beds of grey marlstone up to 1.5 m thick | 11.1 | 24.2 |
| Marl, red | 10.6 | 34.8 |
| Marlstone, red and grey | 7.0 | 41.8 |
| Rock, grey and brown | 2.8 | 44.0 |
| Marl, red and grey | 30.1 | 74.1 |
| Marl, red | 12.5 | 86.6 |
| Sandstone, brown, thin marl at base | 2.1 | 88.7 |
| Sandstone, red and grey | 2.7 | 91.4 |
| Marlstone, red and grey, thin marl at base | 2.8 | 94.2 |
| Sandstone, fine, red and green, a thin marl in the middle | 11.0 | 105.2 |

|  | Thickness m | Depth m |
|---|---|---|
| Marlstone, brown, thin sandstone at base | 1.8 | 107.0 |
| Marl, red | 1.8 | 108.8 |
| Sandstone, fine, red | 6.1 | 114.9 |
| Marlstone, red | 4.0 | 118.9 |
| Marl, grey | 3.8 | 122.7 |
| Rock, red, thin red marl in middle | 3.8 | 126.5 |
| Marl, red and grey | 2.1 | 128.6 |
| Marlstone and marl, red | 4.4 | 134.0 |

HELSBY SANDSTONE

| | | |
|---|---|---|
| Sandstone, fine, red | 18.7 | 152.7 |

## Delamere (Blackbank Wood) Pumping Station No. 2 Borehole (56 NE/5)

Surface level 91 m above OD
National Grid reference 5605 6772
Drilled 1952 for Mid and South-east Cheshire Water Board
Cores examined by D. Thomas and J. R. Earp

|  | Thickness m | Depth m |
|---|---|---|
| QUATERNARY |  |  |
| Soil and sand | 0.5 | 0.5 |
| HELSBY SANDSTONE |  |  |
| Sandstone, hard, red | 6.5 | 7.0 |
| Conglomerate with marl at top and at base | 1.0 | 8.0 |
| Sandstone, hard, red and grey, marl layers at 15 m and 17.5 m | 14.0 | 22.0 |
| Sandstone, red, medium-grained, scattered pockets of red mudstone, a red mudstone parting at base | 9.0 | 31.0 |
| Sandstone, red, medium-grained, many pockets of red mudstone at base | 4.0 | 35.0 |
| Sandstone, red, brown or grey-streaked, medium- to coarse-grained, a red marly layer at 41 m, a fine-grained grey micaceous layer at base | 16.0 | 51.0 |
| Sandstone | 1.0 | 52.0 |
| Sandstone, red and brown, medium-grained, a fine grey shaly bed at 54.5 m, red mudstone pockets in lower part, a red micaceous fissile bed at base | 6.0 | 58.0 |
| Sandstone, mainly red, medium-grained, red mudstone pockets near top, at 67.5 m and at base, beds of grey sandstone at 59 m and 66.5 m, a ripple-marked bedding plane at base | 11.0 | 69.0 |
| Sandstone, grey and brown | 3.0 | 72.0 |
| Sandstone, mainly red with scattered pockets of red mudstone, about 0.5 m of massive buff sandstone at base | 4.0 | 76.0 |
| UPPER MOTTLED SANDSTONE |  |  |
| Sandstone, soft, red and buff mottled, cross-bedded, a mudstone parting at 81.5 m, white mottling and grey nodules in lower part | 7.5 | 83.5 |
| Sandstone, soft, red and mottled, upper part shaly | 4.5 | 88.0 |
| Sandstone, harder, red and shaly, light mottling and banding at about 90 m and in lower 3 to 4 m | 12.0 | 100.0 |
| Sandstone, soft, shaly | 20.0 | 120.0 |
| Sandstone, very soft, red with occasional beds and bands of slightly harder grey sandstone | 124.0 | 244.0 |

## Dunham-on-the-Hill Borehole (47 SE/11)

Surface level 17.4 m above OD
National Grid reference 4683 7146
Drilled 1958 for Geological Survey of Great Britain
Cores examined by J. R. Earp

|  | Thickness m | Depth m |
|---|---|---|
| QUATERNARY |  |  |
| Boulder clay | 6.4 | 6.4 |
| Sand, coarse, gravelly | 12.8 | 19.2 |
| PEBBLE BEDS |  |  |
| Sandstone _no core_ | 7.3 | 26.5 |
| Sandstone, reddish brown, medium- to very coarse-grained, a few cross-bedded and 'millet-seed-grained' layers, a fine-grained buff band at 27.9 m, quartz and igneous pebbles in top 1 m, at 29.5 m and with marl pellets at 30.5 to 30.7 m | 6.1 | 32.6 |
| Sandstone, reddish brown, very coarse-grained, medium- to small-sized pebbles at top, at 33.2 m and at bottom, base fairly sharp | 2.6 | 35.2 |
| LOWER MOTTLED SANDSTONE |  |  |
| Sandstone, reddish brown with a few buff bands, fine to medium-grained becoming coarser and with 'millet-seed' grains below 46 m, red mudstone partings at 35.7 m, 37.8 m, 42 m, 42.6 m, 51.2 m, 51.8 m and 57 m, thin 'marl-pellet' layers at 46.8 m and 57.7 m | 22.9 | 58.1 |
| Mudstone, soft, red with layers and lenses up to 50 cm thick of grey-green to reddish sandstone, 8 cm of 'marl-pellet' conglomerate at 61 m | 3.2 | 61.3 |
| Sandstone, reddish brown, medium-grained, a 23 cm layer of soft red mudstone with sandstone lenses at base | 1.9 | 63.0 |
| Sandstone, soft, reddish brown, medium-grained with thin more argillaceous layers, some harder coarser bands below 66.7 m | 5.6 | 68.6 |
| Sandstone, soft, reddish brown, medium-grained | 1.4 | 70.0 |
| Sandstone, soft, reddish brown, mainly coarse and of 'millet-seed' texture, bedding mainly indistinct but probably cross-bedded at 83 to 84 m and visibly so at around 141 m, very soft and incoherent layers at around 81 m, 86 m, 90 m, 107.5 m, 119.5–120 m and 124–124.5 m | 76.6 | 140.6 |
| Sandstone, reddish-brown, coarse-grained, well laminated, bottom 15 cm fissured | 4.6 | 151.2 |

## Eaton Pumping Station No. 2 Borehole (56 SE/4)

Surface level about 85 m above OD
National Grid reference 5672 6341
Drilled 1939 for Crewe Corporation Waterworks
Cores examined by D. A. Wray

|  | Thickness m | Depth m |
|---|---|---|
| QUATERNARY |  |  |
| Reddish stony loam | 3.0 | 3.0 |
| TARPORLEY SILTSTONES |  |  |
| Sandstone, hard, 'marly', a strongly ripple-marked surface with four footprints of a three-toed reptile | 0.7 | 3.7 |

|  | Thickness m | Depth m |
|---|---|---|
| Marl, chocolate with bands of hard 'marly' sandstone up to 30 cm, several ripple-marked surfaces | 16.3 | 20.0 |
| Sandstone, chocolate, 'marly', ripple-marked, with bands of chocolate 'marl' up to 30 m greenish grey 'marly' sandstone bands towards base, salt pseudomorphs at 31 m | 12.5 | 32.5 |
| Sandstone, mainly coarse, reddish, thin 'marly' bands and ripple marks, a 30 cm light greenish mottled bed at base | 3.0 | 35.5 |

HELSBY SANDSTONE

|  | Thickness m | Depth m |
|---|---|---|
| Sandstone, brownish-red, fine to medium, rounded-grained, cross-bedded | 10.0 | 45.5 |
| Sandstone, harder, fine-grained, top 46 cm grey-green mottled, on chocolate sandstone with marly partings | 3.5 | 49.0 |
| Sandstone, soft, reddish brown, coarse-grained, cross-bedded, concretions of barytes, some 60 cm of fine-grained chocolate sandstone at base | 8.3 | 57.3 |
| Sandstone, purple and red, 'marly', with green bands | 3.0 | 60.3 |
| Sandstone, fine-grained, grey, plant stems up to 3 cm wide, 60 cm long, one branched | 0.7 | 61.0 |
| Sandstone, purplish, grey bands and 'marl' pellets at top, 'marly' bands at 61.25–62.25, quartz pebbles at 62.5 | 2.5 | 63.5 |
| Sandstone, brown or purplish with green or grey mottling or banding, top 2 m with quartz pebbles and 'marl' pellets resting on an eroded surface, sandstones coarser, marly and yellowish towards base | 7.5 | 71.0 |
| 'Marl' purple with green mottling. | 2.0 | 73.0 |
| Sandstone, fine-grained, purplish with green mottled 'marl' partings, one at 74.5 m showing sun cracks and burrows | 2.0 | 75.0 |
| Sandstone, medium-grained, purplish with green micaceous bands, small quartz pebbles in basal 45 cm, base irregular on a green 'marl' parting | 5.5 | 80.5 |
| Sandstone, medium-grained, purplish brown with green bands or grey patches, a sandy marl band 15 cm at 83 m, basal 30 cm coarse and pebbly | 6.0 | 86.5 |
| Sandstone, medium- to coarse-grained, purplish brown, green or grey mottled, upper 12 m with 'marl' partings, lower 12 m with quartz pebbles up to 4 cm and 'marl' fragments, a layer of large 'marl' fragments at 109.5 m | 24.5 | 111.0 |
| Sandstone, red and grey, medium- to coarse-grained, many 'marl' pellets below 112.5 m, abundant at 113.5 m and at base | 5.0 | 116.0 |
| 'Marl', chocolate and green, alternating with 'marly' sandstone bands | 1.5 | 117.5 |
| Sandstone, red-brown, with 'marl' partings and pellets | 1.5 | 119.0 |
| Sandstone, red-brown, medium-grained, some white, ?quartzitic joints above 122 m, occasional greenish grey bands and 'marl' pellets below | 9.0 | 128.0 |
| Sandstone, fine-grained, 'marly', upper surface ripple-marked, many sun cracks and burrows | 4.6 | 132.6 |

UPPER MOTTLED SANDSTONE

|  |  | Thickness m | Depth m |
|---|---|---|---|
| Sandstone, incoherent, reddish, changing to foxy red at 135.5 m | no core | 22.4 | 155.0 |
| Sandstone, grey-green passing to red-brown, fine-grained, 'marly' lenses, a 'marl' bed 30 cm at 157 m, rare 'marl' pellets below |  | 6.0 | 161.0 |
| Sandstone, coarse, brown, passing to yellow and grey, many large 'marl' pellets and quartz pebbles up to 3 cm |  | 3.5 | 164.5 |
| Sandstone, soft, coarse, bright red, upper part cross-bedded with 'millet-seed' grains, lower part with grey-green bands |  | 6.0 | 170.5 |
| Sandstone, harder, reddish brown, medium-grained, basal 30 cm crowded with chocolate 'marl' pellets |  | 3.5 | 174.0 |
| Sandstone, alternating reddish brown and greyish green, even-grained, rare 'marl' pellets |  | 9.0 | 183.0 |
| Sandstone, soft, foxy red | no core | 5.0 | 188.0 |
| Sandstone, red and grey, flaggy |  | 5.0 | 193.0 |
| Sandstone, incoherent, foxy red and grey | no core | 6.5 | 199.5 |
| Sandstone, red and mottled, flaggy |  | 1.5 | 201.0 |
| Sandstone, incoherent, red | no core | 18.5 | 219.5 |
| Sandstone, soft, red and grey |  | 24.5 | 244.0 |

**Eddisbury Pumping Station Borehole** (56 NE/1)
Surface level 102 m above OD
National Grid reference 5578 6943
Drilled 1935 and 1939 for Northwich Rural District Council
Cores to 153 m examined by W. B. Wright

|  | Thickness m | Depth m |
|---|---|---|

QUATERNARY

|  | Thickness m | Depth m |
|---|---|---|
| Made ground and sand | 2.0 | 2.0 |

HELSBY SANDSTONE

|  | Thickness m | Depth m |
|---|---|---|
| Sandstone, soft, red, passing to grey mottled, a hard rib at 3.5 m | 7.0 | 9.0 |
| Sandstone, red, yellow and grey with micaceous marly bands | 3.0 | 12.0 |
| Sandstone, mainly grey and red | 6.0 | 18.0 |
| Sandstone, red, grey and yellow mottled, marl pockets and quartz pebbles at 22–23 m and near base | 7.5 | 25.5 |
| Sandstone, dark red, a thin breccia at 26 m, a marly layer at base | 4.5 | 30.0 |
| Sandstone, red and mottled with marly layers | 3.5 | 33.5 |
| Sandstone, hard red and grey, with marl pockets and quartz pebbles | 5.5 | 39.0 |
| Sandstone, red and grey with layers of marl | 3.0 | 42.0 |
| Sandstone, red and grey with marl-conglomerates at 44 m and at base | 4.5 | 46.5 |
| Sandstone, mainly soft, red, the bottom 2 m harder | 12.5 | 59.0 |
| Sandstone, red and grey, fine to coarse, mainly fairly soft | 11.0 | 70.0 |
| Sandstone hard, red, laminated marly bands, marl pebbles at base | 7.0 | 77.0 |
| Sandstone, coarse grey | 0.5 | 77.5 |

UPPER MOTTLED SANDSTONE

|  | Thickness m | Depth m |
|---|---|---|
| Sandstone, fine, red, traces of marl | 2.5 | 80.0 |
| Sandstone, red, alternating hard and soft | 2.0 | 82.0 |

| | Thickness m | Depth m |
|---|---|---|
| Sandstone, red, very soft | 14.0 | 96.0 |
| Sandstone, red, traces of marl, a 3 cm marl parting at base | 11.0 | 107.0 |
| Sandstone, soft, red, white or red concretions at 115 m, 117 m, 122.5 m, 124 m and 147 m | 46.0 | 153.0 |
| Sandstone, soft, red, with occasional grey, or yellowish beds | 76.0 | 229.0 |

## Foxhill Pumping Station No. 1 (Main) Borehole (57 SW/1)
Surface level 48 m above OD
National Grid reference 5064 7484
Drilled 1939 for Warrington Corporation Water Works Dept.

| | Thickness m | Depth m |
|---|---|---|
| QUATERNARY | | |
| Soil, clay and sand | 1.8 | 1.8 |
| Sand and gravel | 7.3 | 9.1 |
| UPPER MOTTLED SANDSTONE | | |
| Sandstone, mainly soft, red | 39.7 | 48.8 |
| Sandstone, red and grey | 43.9 | 92.7 |
| Sandstone, soft red | 3.6 | 96.3 |
| Sandstone, broken | 3.1 | 99.4 |
| Sandstone, soft red | 21.6 | 121.0 |
| Sandstone, broken | 31.4 | 152.4 |
| Sandstone, soft red | 39.3 | 191.7 |
| Sandstone, soft grey | 21.4 | 213.4 |
| Sandstone, red and grey in thin layers | 13.7 | 227.1 |
| Sandstone, mainly grey | 48.0 | 275.1 |

## Foxhill Trial Borehole (57 SW/4)
Surface level 64 m above OD
National Grid reference 5111 7483
Drilled 1937 for Warrington Corporation Waterworks Dept.

| | Thickness m | Depth m |
|---|---|---|
| QUATERNARY | | |
| Soil and clayey sand | 4.3 | 4.3 |
| Sand, clay and stones | 7.6 | 11.9 |
| UPPER MOTTLED SANDSTONE | | |
| Hard sand and sandy marl | 14.6 | 26.5 |
| Sandstone, red | 12.4 | 38.9 |
| Sandy marl | 1.4 | 40.3 |
| Sandstone, soft red | 40.4 | 80.7 |
| Sandy marl and sandstone | 8.4 | 89.1 |
| Sandstone, red and white | 18.6 | 107.7 |
| Sandstone, red, brown and grey | 11.1 | 118.8 |
| Sandstone, soft red | 6.0 | 124.8 |
| Sandstone, red and white | 12.5 | 137.3 |
| Sandstone, red | 16.7 | 154.0 |

## Handley Observation Borehole (45 NE/26)
Surface level about 22 m above OD
National Grid reference 4678 5748
Drilled 1973 for Dee and Clwyd River Authority
Cores examined by T. J. Charsley and R. J. Follows

| | Thickness m | Depth m |
|---|---|---|
| No record | 13.50 | 13.50 |
| LOWER MOTTLED SANDSTONE | | |
| Sandstone, red-brown, fine- to coarse- | | |

| | Thickness m | Depth m |
|---|---|---|
| grained with silty bands, strongly cross-bedded, some black spotting | 14.50 | 28.00 |
| Sandstone, orange-brown, mainly fine-grained with coarse bands | 2.75 | 30.75 |
| Sandstone, orange-brown becoming red-brown downwards, greenish grey bands at around 50 m and spotting or mottling from 61 to 73 m, mainly medium-grained with thin layers alternating from coarse to silty, mainly cross-bedded | 56.25 | 87.00 |
| Sandstone, red-brown, fine- to medium-grained occasional cross-bedded silty horizons, jointed, slickensided and faulted between 90 and 95.5 m | 15.25 | 102.25 |
| Sandstone, red-brown, fine- to medium-grained, mainly hard and with veins and bands of silty sandstone | 14.25 | 116.50 |
| Sandstone, red-brown, fine- to medium-grained, mostly slickensided and broken | 2.00 | 118.50 |
| Sandstone, red-brown, fine- to medium-grained and mainly silty, well bedded to thinly bedded and in places fissile | 25.50 | 144.00 |
| Sandstone, red-brown, fine- to medium-grained but somewhat silty and massive, fine yellow spotting common | 15.00 | 159.00 |

## Hondslough Farm Borehole (57 SW/19)
Surface level 94 m above OD
National Grid reference 5370 7258
Drilled 1958 for Geological Survey of Great Britain
Cores examined by J. R. Earp and B. J. Taylor

| | Thickness m | Depth m |
|---|---|---|
| QUATERNARY | | |
| Sand | 2.0 | 2.0 |
| Sand and gravel | 5.0 | 7.0 |
| TARPORLEY SILTSTONES | | |
| Mudstone, red, grey and green | 3.0 | 10.0 |
| Siltstone and sandstone, grey-green | 4.0 | 14.0 |
| Mudstone and shale, red, brown and green, passing down into a grey and brown siltstone and sandstone, a thin sandstone and conglomerate at base | 4.0 | 18.0 |
| Mudstone and siltstone, brown and grey, top 2 m autobrecciated | 2.5 | 20.5 |
| Sandstone, fine- to medium-grained, brown with light and dark mottling | 6.5 | 27 |
| Shale, red, brown and green banded, with micaceous sandy laminae | 0.4 | 27.4 |
| Sandstone, mainly fine- to medium-grained, brown with light and dark mottling | 1.0 | 28.4 |
| Shale, red, brown and green banded, micaceous sandy laminae, ripple marks, etc. | 1.6 | 30.0 |
| Sandstone, medium- to fine-grained, with silty layers, brown, buff or with grey-green banding, green shaly laminae towards base | 7.8 | 37.8 |
| Shale, with silty laminae, brown, mauve, green and grey banded | 0.7 | 38.5 |
| Sandstone, medium-grained, brown and grey, green banded at top and base | 5.0 | 43.5 |
| Shale, red, mauve and green banded, micaceous sandy laminae, salt pseudomorphs at 44 m | 1.5 | 45.0 |

| | Thickness m | Depth m |
|---|---|---|
| Sandstone, mauve and brown coarsening downwards from fine to coarse | 1.0 | 46.0 |
| Shale, red and green, micaceous sandstone, brown, medium- to coarse-grained | 0.6 | 46.9 |
| Shale, red and green, micaceous sandy laminae | 0.7 | 47.6 |
| Sandstone, brown, medium-grained, top and bottom shaly | 1.4 | 49.0 |
| Shale, red and green, micaceous sandy laminae; a thin brown sandstone in middle | 0.6 | 49.6 |
| Sandstone, brown, medium- to fine-grained and silty | 3.8 | 53.4 |
| Mudstone, chocolate and grey, micaceous with silty layers and sandy at base salt pseudomorphs at 54.2 m | | |
| Sandstone, red and grey, medium-grained, turbulent bedding in upper and lower parts | 1.8 | 56.6 |
| Mudstone, red and grey, with paler silty and sandy laminae | 4.2 | 60.8 |
| Alternations of red sandstone and 'striped' mudstone/siltstone beds, small fault at 62.5 m, large salt pseudomorphs at 66.5 m | 8.2 | 69.0 |
| Sandstone 'mush' and red sandstone | 2.0 | 71.0 |
| Alternations of red sandstone and 'striped' mudstone/siltstone beds, large salt pseudomorphs at 74 m a small fault at c. 74.5 m, mud cracks at 78.5 m | 8.5 | 79.5 |
| Alternations of red shale and buff siltstone, many micaceous layers, grey-green layers towards base | 3.0 | 82.5 |
| Sandstone, brown-grey, fine- to medium-grained, well bedded | 2.5 | 85.0 |

HELSBY SANDSTONE

| | Thickness m | Depth m |
|---|---|---|
| Sandstone, reddish brown, medium- to coarse-grained mainly incoherent | 28.1 | 113.1 |
| Sandstone, reddish brown, fine-grained, passing down into siltstone and red mudstone, a fissure at base | 1.9 | 115.0 |
| Sandstone, brown, medium- to coarse-grained, scattered clay galls, green shaly partings at 119.5 m, coarse and pebbly at base | 6.0 | 121.0 |
| Conglomerate and sandstone, with clay galls and a quartz pebble | 1.0 | 122.0 |
| Sandstone, medium- to coarse-grained, brown, lower part much fissured | 6.5 | 128.5 |
| Sandstone, coarse, brown, passing down into very coarse cross-bedded conglomerate | 2.7 | 131.2 |
| Sandstone, brown and grey, fine- to medium-grained with micaceous laminae | 0.3 | 131.5 |
| Sandstone, brown with lighter brown inclusions, passing down into conglomerate | 2.3 | 133.8 |

**Meadowbank Mine No. 3 Shaft** (66 NE/4)
Surface level 26.86 m above OD
National Grid reference 6525 6825
Log by Imperial Chemical Industries, Ltd.

| | Thickness m | Depth m |
|---|---|---|
| Drift | 2.44 | 2.44 |
| Grey marl | 1.75 | 4.19 |
| Red clay | 1.83 | 6.02 |

| | Thickness m | Depth m |
|---|---|---|
| Red and grey marl | 6.71 | 12.73 |
| Grey marl | 14.55 | 27.28 |
| Red and grey marl with gypsum | 20.57 | 47.85 |
| Red and grey shaly marl | 10.67 | 58.52 |
| Hard blue marlstone | 0.61 | 59.13 |
| Hard red and grey marl | 1.83 | 60.96 |
| Rock salt | 6.10 | 67.06 |
| Hard red and blue marl | 3.04 | 70.10 |
| Rock salt | 3.36 | 73.46 |
| Hard red and grey marl | 0.91 | 74.37 |
| Rock salt | 7.32 | 81.69 |
| Rock marlstone | 0.61 | 82.30 |
| Rock salt (Top Bed) | 29.26 | 111.56 |
| Red and grey marl with rock salt | 10.36 | 121.92 |
| Rock salt (Bottom Bed) | 25.60 | 147.52 |

**Esso Petroleum, Milton Green No. 1 Borehole** (45 NW/9)
Surface level 16 m above OD
National Grid reference 4374 5692
Drilled 1965 for Esso Petroleum Company
Material from cored horizons examined for British Geological Survey by D. Price and R. K. Harrison

| | Thickness m | Depth m |
|---|---|---|
| QUATERNARY | | |
| Unrecorded | 10 | 10 |
| Clay and sand | 19 | 29 |
| UPPER COAL MEASURES | | |
| Siltstone and clay, red-brown to grey-green | 37 | 66 |
| Claystone, red-brown to grey-green; sandy at about 113–115 m | 78 | 144 |
| Siltstone, red-brown and grey; coaly bands in lower 8 m | 28 | 172 |
| Siltstone, sandy, and sandstone, red-purple to grey-green | 43 | 215 |
| Claystone, red-brown to grey-green; coaly bands in lower 12 m | 41 | 256 |
| Siltstone and sandstone, grey to red-brown; claystones and coaly bands in lower 15 m | 63 | 319 |
| Sandstone and siltstone, mainly grey-green; coal seams at 355 to 360 m | 79 | 398 |
| Shale, red-brown to grey-green, and siltstone; thin coal in bottom 1 m | 57 | 455 |
| Claystone, red-brown to grey-green | 72 | 527 |
| MIDDLE AND LOWER COAL MEASURES | | |
| Claystone and shale, grey with many coal horizons | 47 | 574 |
| Siltstone and sandstone, grey; lower part shaly with a coal seam | 39 | 613 |
| Siltstone and shale, grey with many coal seams; some possible correlatives include the Upper Stinking at 651 m, Main at 763 m and Fireclay Group at the base | 204 | 817 |
| Siltstone, grey; some shale and sandstone layers; thin coals | 89 | 906 |
| Sandstone, white, passing down into grey siltstone and silty shale. thin coals in the lower 27 m | 104 | 1010 |
| MILLSTONE GRIT SERIES | | |
| Sandstone, white, medium- to coarse-grained, silaceous | 17 | 1027 |

| | Thickness m | Depth m |
|---|---|---|
| Siltstone and shale, grey with thin sandstones | 111 | 1138 |
| Sandstone, white, with grey shaly layers | 40 | 1178 |
| **CARBONIFEROUS LIMESTONE SERIES** | | |
| Limestone, cream to brown slightly sandy | 5 | 1183 |
| Limestone, dark grey, argillaceous and fossiliferous | 97 | 1280 |
| Sandstone, brownish, fine- to medium-grained, very calcareous | 3 | 1283 |
| Shale, dark to very dark grey, silty and calcareous | 14 | 1297 |
| Limestone, cream to brown, slightly argillaceous and in parts dolomitic | 61 | 1358 |
| Limestone, brown, argillaceous and fossiliferous | 15 | 1373 |
| Limestone, white to light grey-brown, parts dolomitised, fissile | 75 | 1448 |
| Dolomite, light grey, bottom 7 m or so with dolomitised limestone | 22 | 1470 |
| Limestone, white to brown | 30 | 1500 |
| Shale-limestone and red, silty, calcareous sandstone | 2 | 1502 |
| Silty sandstone and sandy siltstone, grey-brown | 65 | 1567 |
| **ORDOVICIAN** | | |
| Mudstone, greenish grey, chloritic, striped and silty with lineated bedding planes; about 1 m of highly altered igneous rock at 1586 m | 11 | 1588 |

**Mouldsworth Pumping Station, No. 2 Borehole** (57 SW/17)
Surface level 44 m above OD
National Grid reference 5028 7036
Drilled 1932–1934 for West Cheshire Water Board

| | Thickness m | Depth m |
|---|---|---|
| **QUATERNARY** | | |
| Soil and boulder clay | 1.1 | 1.1 |
| Marl, red with blue veins | 4.0 | 5.1 |
| Sand, thin gravel at base | 27.4 | 32.5 |
| Marly sand | 5.0 | 37.5 |
| Gravel and mainly coarse sand | 38.4 | 75.9 |
| **UPPER MOTTLED SANDSTONE** | | |
| Sandstone, soft, red, loamy and laminated | 0.8 | 76.7 |
| Sandstone, red, grey bands in lower part | 8.7 | 85.4 |
| Sandstone, red, fissured | 4.3 | 89.7 |
| Sandstone, red, with loamy beds and grey sands | 45.9 | 135.6 |
| **PEBBLE BEDS** | | |
| Sandstone, red, occasional grey bands and scattered pebbles | 50.6 | 186.2 |
| Sandstone, red with pebbles | 29.6 | 215.8 |
| Sandstone, red, medium-grained | 14.0 | 229.8 |
| Sandstone, red, coarser-grained with pebbles | 15.9 | 245.1 |
| Sandstone, red, mainly coarse | 26.6 | 271.7 |
| Sandstone, red with pebbles | 35.3 | 307.0 |

**Newton Hollow, Ravenslodge Trial Borehole** (57 SW/11)
Surface level 113 m above OD
National Grid reference 5314 7412
Drilled 1937 for Warrington Corporation Waterworks Dept.
Cores examined by W. B. Wright, R. C. B. Jones and
D. A. Wray

| | Thickness m | Depth m |
|---|---|---|
| **TARPORLEY SILTSTONES** | | |
| Marl, chocolate with grey-green bands and beds of chocolate sandstone | 62.5 | 62.5 |
| Marl, sandy, chocolate, alternating with marly sandstones | 13.7 | 76.2 |
| Sandstone, fine, red | 5.2 | 81.4 |
| Sandstone, platy with marl streaks | 3.3 | 84.7 |
| Sandstone, fine-grained, red | 7.0 | 91.7 |
| Sandstone, banded, chocolate, marly | 1.9 | 93.6 |
| Sandstone, fine, red | 2.7 | 96.3 |
| Sandstone, platy, chocolate, marly | 2.2 | 98.5 |
| Sandstone, red with grey-green marly bands | 11.2 | 109.7 |
| Marl, shaly, red and grey-green with red sandstone bands up to 0.5 m thick | 22.6 | 132.3 |
| **HELSBY SANDSTONE (FRODSHAM BEDS FACIES)** | | |
| Sandstone, soft, red, mottled white and yellow, harder towards top | 17.1 | 149.4 |

**Newton Hollow Pumping Station, No. 3 Borehole** (57 SW/7)
Surface level 134 m above OD
National Grid reference 5258 7420
Drilled 1946 for Warrington Corporation Waterworks Dept.
Some core seen by A. W. Allen

| | Thickness m | Depth m |
|---|---|---|
| **TARPORLEY SILTSTONES** | | |
| Marl, hard, chocolate | 4.9 | 4.9 |
| **HELSBY SANDSTONE** | | |
| Sandstone, red and brown | 6.4 | 13.3 |
| Sandstone, grey | 0.6 | 13.9 |
| Sandstone, soft, red, occasional marl nodules | 16.3 | 30.2 |
| Sandstone, harder, red, mottled with grey-green patches | 2.7 | 32.9 |
| Sandstone, red | 6.7 | 39.6 |
| Sandstone, dark red with marly layers and pockets | 11.8 | 51.4 |
| Sandstone, mottled, with micaceous shaly partings | 0.4 | 51.8 |
| Sandstone, red | 4.5 | 56.3 |
| Sandstone, mottled grey-green, micaceous partings | 2.2 | 58.5 |
| Sandstone, dark brown, many shaly partings | 4.3 | 62.8 |
| Sandstone, red | 16.1 | 78.9 |
| Sandstone, red with thin conglomerates | 3.4 | 82.3 |
| Sandstone, red | 3.0 | 85.3 |
| Sandstone, red, bands of marl at top | 4.9 | 90.2 |
| Marl | 0.6 | 90.8 |
| Sandstone, red, pockets of marl near base | 5.5 | 96.3 |
| **UPPER MOTTLED SANDSTONE** | | |
| Sandstone, soft, bright red, cross-bedded with patches of marl | 18.8 | 115.1 |
| Marl | 2.9 | 118.0 |
| Sandstone, soft, bright red, some layers grey mottled, very small pebbles between 183 m and 192 m | 108.6 | 226.6 |
| Sandstone, soft, red | 78.5 | 305.1 |

**Oakmere Borehole** (53 NE/14)
Surface level 81.19 m above OD
National Grid reference 5768 6780
Drilled 1963 for Cheshire Brine Subsidence Compensation Board
Log by B. J. Taylor, from core examination to 186 m, from drillers log and gamma log to base

| | Thickness m | Depth m |
|---|---|---|
| QUATERNARY | | |
| Sand | 38.10 | 38.10 |
| Boulder Clay, incorporating mudstone debris near base | 4.57 | 42.67 |
| MERCIA MUDSTONE GROUP | | |
| Mudstone, red and grey, alternately finely banded and blocky; beds horizontal near top, dip 8° at 86 m. Gypsum veins throughout | 49.68 | 92.35 |
| Mudstone as above, autobrecciated in places (? by collapse) gypsum in veins and large nodules | 10.06 | 102.41 |
| Mudstone, red and grey, alternately banded and blocky; gypsum veins and nodules | 16.16 | 118.57 |
| NORTHWICH HALITE FORMATION | | |
| Breccia of red and grey mudstone | 27.12 | 145.69 |
| Mudstone, red and grey; banded at intervals; gypsum veins and nodules | 10.98 | 156.67 |
| Rock salt, orange, wisps of mudstone | 0.43 | 157.10 |
| Mudstone, brown; salt crystals | 0.91 | 158.01 |
| Rock salt, orange; muddy at intervals; small gypsum inclusions | 23.42 | 181.43 |
| Mudstone, with beds of muddy salt | 15.32 | 196.75 |
| Mudstone, with salt veins | 10.21 | 206.96 |
| Mudstone, with gypsum veins | 3.35 | 210.31 |
| Mudstone, red and grey | 3.35 | 213.66 |

**Old Beachin Farm Borehole** (45 NW/8)
Surface level 11.8 m above OD
National Grid reference 4452 5732
Drilled 1958 for Geological Survey of Great Britain
Cores examined by J. R. Earp and W. B. Evans

| | Thickness m | Depth m |
|---|---|---|
| QUATERNARY | | |
| Mainly reddish brown clay with stones | 14.2 | 14.2 |
| UPPER COAL MEASURES | | |
| Mudstone alternating with sandy mudstone, red with green banding or mottling, occasional 'fish-eyes', polished quartz grains, concretionary layers, a few striped siltstones, quartz pebbles at about 54–56 m, possible plant at 29.9 m, mussel at 45 m and 64 m | 50.4 | 64.6 |
| Mudstone, mainly red, slickensided, *Spirorbis* at 64.8 m, plant at 65.2 m, ostracods, plants etc. at 65.8 to 66.2 m | 1.6 | 66.2 |
| Mudstone and sandy mudstone, red and green, possible fossils at 69 m and 70.4 m | 6.3 | 72.5 |
| Sandstone, pink, silty, striped, a thin red mudstone band at 76.7 m | 4.5 | 77.0 |
| Mudstone, mainly red and silty, sandy or striped, organic debris in top 2.5 m, pebbles at 81 to 82 m | 11.2 | 88.2 |
| Conglomerate, with much ironstone | 1.4 | 89.6 |

| | Thickness m | Depth m |
|---|---|---|
| Mudstone, mainly sandy or striped, silty, mainly purple-red, 2.2 m of striped sandstone at 95.2 m, plant debris at about 97 to 100 m | 11.9 | 101.5 |
| Sandstone, mainly pale red, silty and striped, and red sandy mudstone, plant debris down to 104.3 m, bottom 1.2 m grey-green, banded | 7.8 | 109.3 |
| Mudstone, and siltstone, mainly purplish with green markings below 114 m, pebbles in top 1.6 m, plant fragments at 112.7 m and between 115.5 and 117 m | 10.7 | 120.0 |
| Sandstone and siltstone, massive mauve-brown | 0.4 | 120.4 |
| Mudstone and shale, purplish with green, micaceous, silty laminae towards base | 0.7 | 121.1 |
| Sandstone and siltstone, tough, red-brown, laminated, micaceous towards base | 1.1 | 122.2 |
| Mudstone, soft, purplish red, partly silty with green vermicular markings, 'fish-eyes' and pellety granules, large plant stems at around 126 and 136 m, 46 cm of tough siltstone at 127.2 m, 5 cm of conglomerate at 128.9 m and 1 m of siltstone at 137 m | 20.3 | 142.5 |
| Sandstone, fine- to medium-grained, mauve, grey and brown, a thin mottled siltstone band at base | 1.0 | 143.5 |
| Mudstone, purplish red and green, shaly layers, bottom 46 cm silty | 4.6 | 148.1 |
| Siltstone, tough, purple to mauve to grey, a plan-bearing mudstone layer 50 cm on laminated sandstone 30 cm at 153.6 m, sandy layers and ironstone concretions towards base | 8.6 | 156.7 |
| Mudstone, silty at top, variegated from purple to green and ochre, lower part shaly with plants, base fissile sheared | 2.0 | 158.7 |
| Seatearth, grey, carbonaceous | 0.3 | 159.0 |
| Seatearth, grey, lower part soft, ochrous | 2.5 | 161.5 |
| Mudstone, purplish red to grey-green, texture from blocky, silty to shaly or of soft seatearth-type, sandstone layers 25 cm at 164.9 m, 20 cm at 168.7 m, scattered plants at 170 m conspicuous pellety layers at 172.7 m | 11.3 | 172.8 |
| Siltstone and silty mudstone, red, brown with green markings, sandy layers 8 cm at top and 15 cm at base | 1.5 | 174.3 |
| Mudstone, mainly silty, purplish red with green markings, lower 1.5 m soft shaly and with abundant plants | 6.7 | 181.0 |
| Seatearth-mudstone 15 cm, on mudstone, mainly purple-red with green markings, pellety layers, scattered plants | 3.0 | 184.0 |
| Mudstone, silty mudstone and siltstone, purplish with green markings, grey streaks etc., sandy layer with plant-strewn laminae at 189.5 m, plants at around 191.5 and 192.5 m, thin buff sandstones at 194.5 and 195.5 m | 12.3 | 196.3 |
| Mudstone, purple-mauve with green markings, passing down into mauve-grey shaly mudstone, sheared towards base, abundant plants at around 197 m and towards a 10 cm carbonaceous layer at base | 3.7 | 200.0 |

| | Thickness m | Depth m |
|---|---|---|
| Seatearth, grey, with purple and yellow veins | 1.0 | 201.0 |
| Mudstone, green to purple, roots in upper part, sandy with ironstone and limestone nodules towards base | 3.7 | 204.7 |
| Siltstone and gritty sandstone, grey, with 90 cm red-brown sandy mudstone in middle | 2.3 | 207.0 |
| Pseudobreccia of red-brown mudstone and grey sandstone | 1.2 | 208.2 |
| Mudstone, red-brown, upper 1.4 m with grey silty banding, lower part mainly blocky | 4.8 | 213.0 |
| Sandstone and siltstone, striped, grey with partings of chocolate mudstone | 4.9 | 217.9 |
| Mudstone, grey, shaly, red and yellow ironstone bands, worm casts, some coalified plants and *Spirorbis* near base | 1.1 | 219.0 |
| Sandy mudstone and striped sandstone, grey | 1.2 | 220.2 |
| Mudstone, chocolate, plant debris, top part sheared, middle part with disturbed pieces of sandstone and siltstone | 3.3 | 223.5 |
| Mudstone, purple to grey, silty       *only 35% cored* | 5.0 | 228.5 |
| Mudstone, purple-brown, or banded with green, silty at top and bottom | 4.7 | 233.2 |
| Siltstone and silty mudstone, purple to grey-green, mainly striped, plant stems in lower part | 4.4 | 237.0 |
| Mudstone, upper part green to purple, and partly of seatearth type, lower 1.6 m mainly grey with plants, a layer of coalified plant debris at base | 5.0 | 242.6 |
| Seatearth and plant-bearing mudstone, grey, an ironstone and a siltstone layer at base | 1.4 | 244.0 |
| Mudstone, mainly silty and siltstone, grey, laminated, plants | 1.0 | 245.0 |
| Mudstone, grey, shaly, ferruginous bands, bottom 30 cm with plants | 2.0 | 247.0 |
| Seatearth 84 cm, on mainly silty mudstone, grey-green but lowest metre or so purplish, a carbonaceous parting at base | 4.7 | 251.7 |
| Seatearth 70 m on silty mudstone, purple-brown with green layers and patches | 5.3 | 257.0 |
| Mudstone, purple-brown to green, top part with calcareous concretions and lenses, a tough, laminated, silty bed 75 cm at 264 m | 8.5 | 265.5 |
| Mudstone, shaly, variegated, large plant stems on seatearth 8 cm at base | 0.5 | 266.0 |
| Mudstone and silty mudstone, purple to grey-green, calcareous nodules lenses and tubes below 267.5 m, siltstone beds 4.5 cm at 270 m and 90 cm at 273.5 m, plant stems towards base | 12.4 | 278.4 |
| Mudstone, mauve to grey, plant debris, a coaly dirt band 15 cm on dark grey shale with *Euestheria* etc. 23 cm at base | 1.1 | 279.5 |
| Sandstone, medium to coarse, mainly mauve to purple-grey, occasional highly micaceous laminae, layers with mudstone pebbles at 281 m and below 291.5 m, plant-bearing mudstone parting at 293.3 m, bottom 80 cm coarse and pebbly, base sharp, undisturbed | 14.6 | 294.1 |
| Mudstone, mainly purple-brown with green silty layers and green patches, a carbonaceous parting at 296.6 m, thin seatearth at 299.1 m, bottom 1.5 m with greenish siltstones | 6.6 | 300.7 |

| | Thickness m | Depth m |
|---|---|---|
| Sandstone, tough, silty, greenish and mauve | 1.1 | 301.8 |
| Mudstone, mottled purple-brown and green | 1.0 | 302.8 |

**Over Borehole** (66 NW/8)
Surface level 53.34 m above OD
National Grid reference 6364 6587
Drilled 1958 for Geological Survey of Great Britain
Log by B. J. Taylor

| | Thickness m | Depth m |
|---|---|---|
| *Rock-bit drilling* | | |
| QUATERNARY | | |
| Sand, reddish orange; a few brown clay fragments near top | 11.89 | 11.89 |
| Gravel with a little sand; hard rock pebbles and fragments | 6.40 | 18.29 |
| Red clay, stony; abundant mudstone fragments near base | 1.83 | 20.12 |
| MERCIA MUDSTONE GROUP | | |
| Mudstone, red and grey | 4.26 | 24.38 |
| *Core drilling* | | |
| Laminated red and grey mudstone and siltstone, alternating with blocky reddish brown silty mudstone. The laminated facies shows desiccation cracks and consequent penecontemporaneous deformation and ripple marks; the blocky facies shows circular or irregular green spots and mottling and penecontemporaneous brecciation. The whole core is penetrated by fibrous gypsum veinlets | 40.24 | 64.62 |
| Rhythmic alernations as above, but more arenaceous; siltstone layers thicker and more abundant. Gypsum veins throughout. Indeterminate bisaccate miospores, heavily carbonised, at 101.19 m, associated with relatively numerous specimens of *Tasmanites sp.*, a marine microplankton | 45.77 | 110.39 |
| NORTHWICH HALITE | | |
| Breccia of grey and red mudstone, sandy mudstone and siltstone (solution residue) | 8.36 | 118.75 |
| Mudstone, mainly red; salt inclusions | 2.10 | 120.85 |
| Rock salt with mudstone inclusions | 0.41 | 121.26 |
| Mudstone, red and grey; salt inclusions | 2.84 | 124.10 |
| Rock salt, generally muddy; layers of saliferous mudstone | 13.54 | 137.64 |
| Mudstone, salt inclusions | 0.92 | 138.56 |
| Rock salt; middle part of high purity, showing light and dark banding; mudstone inclusions near top and base | 20.57 | 159.13 |
| Mudstone, saliferous, alternating with rock salt | 15.77 | 174.90 |
| Rock salt; mudstone layers in lower half | 14.69 | 189.59 |
| Mudstone, red and grey; salt inclusions | 9.47 | 199.06 |
| Rock salt; high purity between 207 and 218 m with light and dark banding; mudstone inclusions elsewhere | 21.39 | 220.45 |
| Rock salt and saliferous mudstone alternations | 15.21 | 235.66 |
| Rock salt, orange-brown; mudstone inclusions | 15.80 | 251.46 |

| | Thickness m | Depth m |
|---|---|---|
| Mudstone, mainly red; saliferous | 1.07 | 252.53 |
| Rock salt, orange and brown mottled; high purity between 256 and 269 m; mudstone layers and inclusions elsewhere | 29.44 | 281.97 |
| 'Thirty-Foot Marl': red-brown and grey mudstone with salt inclusions; 1 m of muddy salt at 291.59 m | 11.25 | 293.22 |
| 'Bottom Bed': rock salt, mainly high purity | 12.49 | 305.71 |

*General note:* Gypsum traces are general throughout, in rock salt.

## Borehole RM 28 (66 NW/10)

Surface level 38.32 m above OD
National Grid reference 6447 6724
For Imperial Chemical Industries, Ltd.,
Log by courtesy of I.C.I. Ltd.

| | Thickness m | Depth m |
|---|---|---|
| QUATERNARY | | |
| Glacial Drift | 42.37 | 42.37 |
| **Mercia Mudstone Group** | | |
| Red and grey marl | 59.13 | 101.50 |
| NORTHWICH HALITE | | |
| Rock salt with marl bands | 18.90 | 120.40 |
| Rock salt | 19.35 | 139.75 |
| Rock salt with massive marl bands | 37.34 | 177.09 |
| Rock salt | 22.40 | 199.49 |
| Rock salt with marl bands | 25.76 | 225.25 |
| Rock salt (Top Bed) | 26.22 | 251.46 |
| Marl (30 Foot Marl Band) | 10.36 | 261.82 |
| Rock salt (Bottom Bed) | 26.21 | 288.04 |
| LOWER MUDSTONE DIVISION | | |
| Marl with rock salt bands to bottom of borehole | 17.37 | 305.41 |

## Shell, Stanlow No. 2 Borehole (47 SW/9)

Surface level about 4.5 m above OD
National Grid reference 4305 7469
Drilled 1938 for Shell Refining and Marketing Co.

| | Thickness m | Depth m |
|---|---|---|
| QUATERNARY | | |
| Soil and clay | 1.8 | 1.8 |
| Peat | 2.8 | 4.6 |
| Boulder clay | 5.8 | 10.4 |
| Boulder clay with beds of sand | 11.7 | 22.1 |
| Sand and gravel | 4.7 | 26.8 |
| Sand with beds of clay | 3.0 | 29.8 |
| PEBBLE BEDS | | |
| Sandstone, red | 15.9 | 45.7 |
| Sandstone, soft, red with occasional pebbles | 4.6 | 50.3 |
| Sandstone, harder, red with bands of grey sandstone, occasional pebbles, becoming abundant at base | 65.2 | 115.5 |
| Sandstone, red with grey bands | 3.4 | 118.9 |
| Sandstone, very soft, incoherent | 14.0 | 132.9 |
| Sandstone, split and broken | 19.5 | 152.4 |

## Shell, Stanlow No. 19 Borehole (47 SW/13)

Surface level about 4.5 m above OD
National Grid reference 4315 7420
Drilled 1950 for Shell Refining and Marketing Board
Samples examined by R. H. Price and B. J. Taylor

| | Thickness m | Depth m |
|---|---|---|
| QUATERNARY | | |
| Soil and clay | 1.5 | 1.5 |
| Peat | 3.0 | 4.5 |
| Clay with stones | 24.5 | 29.0 |
| Clay and sand | 4.5 | 33.5 |
| PEBBLE BEDS | | |
| Sandstone | 1.4 | 34.9 |
| Sandstone, fine-grained, reddish brown, 'marly' | 2.3 | 37.2 |
| Sandstone, soft, red-brown 'marly' with bands of coarser, angular-grained sandstone | 10.0 | 47.2 |
| Sandstone, hard, red with coarse subangular grains | 2.8 | 50.0 |
| Sandstone, fine, red, 'marly' | 6.4 | 56.4 |
| Sandstone, medium to coarse, subangular-grained, with 'marly' bands | 19.8 | 76.2 |
| Sandstone, hard, red, subangular-grained | 3.0 | 79.2 |
| Sandstone, ill-graded with 'marly' bed | 9.2 | 88.4 |
| Sandstone, soft, red, subangular-grained | 7.6 | 96.0 |
| Sandstone, red, 'marly' | 5.2 | 101.2 |
| Sandstone, soft, red, subangular-grained with pebbles | 20.8 | 122.0 |
| Sandstone, fine-grained, red with rounded and subangular pebbles | 14.9 | 136.9 |
| Sandstone, 'marly', the upper part with angular pebbles | 15.5 | 152.4 |

## Southley Common Borehole (55 NE/6)

Surface level 51.30 m above OD
National Grid reference 5816 5934
Drilled 1958 for Geological Survey of Great Britain
Log by B. J. Taylor

| | Thickness m | Depth m |
|---|---|---|
| *Rock-bit drilling* | | |
| QUATERNARY | | |
| Sand, clayey, with pebbles; probably from sandy till | 4.59 | 4.59 |
| Clay, red-brown slightly calcareous, hard rock fragments | 8.21 | 12.80 |
| Sand, fine clayey; gravelly in places, calcareous (?marine) shell fragments near base | 6.40 | 19.20 |
| Clay, red-brown; pebbles and ?marine shell debris | 1.83 | 21.03 |
| Sand, gravelly near base | 4.57 | 25.60 |
| Gravel; including quartzites and vein quartz | 1.83 | 27.43 |
| Clay, red-brown; gravelly layer near top; Triassic mudstone fragments and erratic pebbles | 8.23 | 35.66 |

*Core drilling*

MERCIA MUDSTONE GROUP
Laminated facies of red and grey mudstone, silty mudstone, with subordinate siltstone layers, alternating with blocky facies of red-brown mudstone with greenish mottlings.

|  | Thickness m | Depth m |
|---|---|---|
| Veinlets of fibrous gypsum throughout; penecontemporaneous deformation, desiccation cracks, show mainly in the laminated facies. Brecciated at base | 79.55 | 115.21 |

NORTHWICH HALITE

| | Thickness m | Depth m |
|---|---|---|
| Breccia of red and grey mudstone and sandy mudstone (solution residue) | 21.86 | 137.07 |
| Rock salt, orange and brown; mudstone in irregular masses and layers | 7.69 | 144.76 |
| Mudstone, red and grey; both blocky and laminated facies, salt inclusions and fibrous veins | 6.42 | 151.18 |
| Rock salt, orange and brown; some clean layers, mudstone inclusions elsewhere | 12.98 | 164.16 |
| Mudstone, red and grey; both blocky and laminated facies, salt in beds, veins and isolated inclusions | 8.43 | 172.59 |
| Rock salt with mudstone inclusions; several thin layers of saliferous mudstone | 9.68 | 182.27 |
| Rock salt, orange and brown; some clean layers | 20.37 | 202.64 |
| 'Thirty-Foot Marl': mudstone, red-brown and grey; blocky at top, otherwise banded; muddy rock salt 0.53 m at 208.84 m | 9.12 | 211.76 |
| Rock salt, orange and brown, mudstone inclusions | 1.90 | 213.66 |
| Mudstone, saliferous; a fibrous salt vein | 0.92 | 214.58 |
| Rock salt, muddy at top | 17.37 | 231.95 |
| Mudstone, red-brown and grey; both laminated and blocky facies present, salt inclusions and saliferous layers at intervals | 19.61 | 251.56 |

## Wardle Borehole (65 NW/1)
Surface level 55 m above OD
National Grid reference 6063 5713
Drilled 1959 for Geological Survey of Great Britain
Log by B. J. Taylor

| | Thickness m | Depth m |
|---|---|---|

*Rock-bit drilling*

QUATERNARY

| | Thickness m | Depth m |
|---|---|---|
| Clay, reddish brown plastic, few stones | 9.14 | 9.14 |
| Sand, clayey and sandy clay | 2.75 | 11.89 |
| Gravel; hard rock boulders | 2.74 | 14.63 |

MERCIA MUDSTONE GROUP

| | Thickness m | Depth m |
|---|---|---|
| Mudstone, red and grey; gypsum traces | 5.49 | 20.12 |

*Core drilling*

| | Thickness m | Depth m |
|---|---|---|
| Mudstone, chiefly red-brown blocky, with green spots and irregular patches; silty mudstone around 23.57 m; many gypsum veins | 36.87 | 56.99 |
| Alternations of blocky mudstone as above, and laminated red and grey mudstone and silty mudstone; the blocky facies predominates; gypsum abundant in veins to 69 m, below this, veins rarer and carry halite, gypsum pellets and grains | 37.45 | 94.44 |
| Alternations of blocky and laminated facies, siltstone layers in the laminated facies becoming more abundant and thicker downwards; desiccation cracks mainly in the | | |

|  | Thickness m | Depth m |
|---|---|---|
| laminated strata; gypsum chiefly as small grains and thin beds, few veins, and rare below 189 m. Miospore assemblage at 183.18 m includes *Tsugaepollenites oriens* Klaus, 1964 suggests late-Anisian to early-Ladinian age | 109.17 | 203.61 |
| Breccia of red and grey mudstone and siltstone, well-cemented (unlike a modern solution residue); fibrous halite veins | 2.69 | 206.30 |
| Mudstone, red and grey, saliferous | 3.71 | 210.01 |

NORTHWICH HALITE

| | Thickness m | Depth m |
|---|---|---|
| Rock salt; inclusions and thin layers of mudstone | 6.60 | 216.61 |
| Mudstone, saliferous | 1.09 | 217.70 |
| Rock salt; inclusions of mudstone; fibrous halite veins | 2.25 | 219.95 |
| Mudstone, brown and grey banded; saliferous | 0.98 | 220.93 |
| Rock salt, orange-brown, muddy; layers of saliferous mudstone with bases at 227 m, 230 m and 237 m | 18.24 | 239.17 |
| Rock salt, orange-brown; few mudstone inclusions, some layers of high purity | 8.48 | 247.65 |

## Willington Lane Exploratory Borehole, Willington (56 NW/16)
Surface level c.4.6 m above OD
National Grid reference 5314 6526
Drilled 1972 for Mid Cheshire Water Board
Cores examined by J. I. Chisholm and T. J. Charsley

| | Thickness m | Depth m |
|---|---|---|

QUATERNARY

| | Thickness m | Depth m |
|---|---|---|
| Soil, clay and sand | 23.8 | 23.8 |

UPPER MOTTLED SANDSTONE

| | Thickness m | Depth m |
|---|---|---|
| Sandstone | 6.7 | 30.5 |
| Sandstone, soft, red, fine- to medium-grained, a silty bed about 0.5 m at 34 m | 4.5 | 35.0 |
| Siltstone, soft, red, with sandy bands and laminae | 37.2 | 72.2 |
| Sandstone, mainly soft, red, fine- to medium-grained, with silty bands and laminae, thicker red-green siltstone beds: 36 cm at 77.75 m, 46 cm at 83.25 m, 46 cm at 92.5 m, 178 cm at 99.75 m, 132 cm at 116 m and 71 cm at 137 m, red mudstone pellets at 78–82 m and at 116 m | 65.0 | 137.2 |
| Sandstone, red, fine- to medium-grained, silty and micaceous laminae | 4.2 | 141.2 |
| Siltstone, mainly red-brown, sandy and micaceous laminae, a silty sandstone bed 0.75 m at 143 m | 3.8 | 145.0 |
| Sandstone, mainly red-brown, occasional green or white layers, fine- to medium-grained but silty at top and at 147.75 m, some micaceous laminae | 19.3 | 164.3 |
| Mudstone, green laminated, micaceous 20 cm, on red-brown micaceous siltstone and mudstone | 0.9 | 165.2 |
| Sandstone, mainly red-brown fine- to medium-grained, green micaceous layers at about 170.25 m, 171.25 m and 175.5 m | 12.7 | 177.9 |

| | Thickness m | Depth m |
|---|---|---|
| Sandstone, red-brown and green, fine-grained, small-scale cross-laminated bands often followed by bands with coarse rounded grains | 1.0 | 178.9 |

PEBBLE BEDS

| | Thickness m | Depth m |
|---|---|---|
| Sandstone, red-brown, fine- to medium-grained, mudstone pebbles at 180 m, pale green micaceous layers at about 182.5 m rare quartz pebbles | 8.7 | 187.6 |
| Sandstone, red-brown, fine- to medium-grained, pale green micaceous silty layers at about 190 m, 195.25 m and 198.5 m, the last with a 2 cm quartz pebble | 10.9 | 198.5 |
| Sandstone, mainly red-brown fine- to medium-grained, rare quartz pebbles in top 3 m and at 213 m, pale green bands at about 202.75 m and 203.25 m | 19.0 | 217.5 |
| Siltstone, grey-green, micaceous 31 cm, on soft red sandstone and grey silty sandstone | 0.8 | 218.3 |
| Sandstone, red, fine- to medium-grained, very soft, a greenish grey micaceous sandy siltstone bed 168 cm at 234 m | 17.6 | 235.9 |
| FAULT? 71 cm of grey-white sand, silt and clay on sandstone, dull red, fine- to medium-grained, with grey or greenish micaceous silty layers at about 238.9 m, 239.75 m and 243.25 m, small siliceous pebbles at about 239 m and in lower part | 7.4 | 243.3 |
| Sandstone, dull red, fine- to medium-grained, a grey laminated, micaceous band 23 cm at 255.5 m, pebbles at top and below 250 m | 21.0 | 264.3 |
| Sandstone, red and grey, pebbly, pebbles up to 6 cm | 3.1 | 267.4 |
| Siltstone, dull red, laminated, sandy | 0.7 | 268.1 |
| Sandstone, dull red, fine- to coarse-grained, pebbles, some exceeding 7 m | 12.9 | 281.0 |
| Mudstone and pebbly sandstone | 1.2 | 282.2 |
| Sandstone, fine- to coarse-grained, scattered pebbles | 8.6 | 290.8 |
| Sandstone, silty, laminated | 2.4 | 293.2 |
| Sandstone, fine- to medium-grained, a greenish silty layer at 299.5 m, pebbles mostly rare except at base | 7.8 | 301.0 |
| Sandstone, dull red, fine- to medium-grained, some silty layers, thin pebbly bands between 302.4 m and 303.2 m | 4.4 | 305.4 |

**Winsford No. 1 Borehole** (66 NE/55)
Surface level 43.26 m above OD
National Grid reference 6541 6511
Drilled 1971 for Cheshire County Council
Log by R. S. Arthurton

| | Thickness m | Depth m |
|---|---|---|
| No cores or samples | 152.4 | 152.4 |

MERCIA MUDSTONE GROUP

| | Thickness m | Depth m |
|---|---|---|
| Mudstone, silty | 5.3 | 157.7 |
| Solution residue | 4.8 | 162.5 |

NORTHWICH HALITE

| | Thickness m | Depth m |
|---|---|---|
| Halite rock, displacive facies in upper part, layered below; red and grey matrix silty | | |

| | Thickness m | Depth m |
|---|---|---|
| mudstone; very muddy in lowest 0.6 m with zoned halite crystals | 2.9 | 168.3 |
| Mudstone, silty, mainly red; displacive halite especially near base | 1.2 | 169.5 |
| Halite rock, mostly displacive but vaguely layered towards top; red silty mudstone matrix | 2.2 | 171.7 |
| Siltstone and silty mudstone, red and grey; load casts; scattered halite crystals | 1.8 | 173.5 |
| Halite rock, muddy; penecontemporaneous-contortion; erosion surface at top | 1.1 | 174.6 |
| Siltstone and silty mudstone, grey; thin layers with halite crystals | 1.0 | 175.6 |
| Halite rock, mostly displacive, penecontemporaneous contortion; erosion surface at top | 0.9 | 176.5 |
| Mudstone, silty, mainly red | 0.3 | 176.8 |
| Halite rock, mostly displacive but layered near base where thin mudstone layer | 4.5 | 181.3 |
| Mudstone, silty red; sparse halite crystals; red and grey towards top where interlaminated halite | 0.8 | 182.1 |
| Halite rock, interspersed layered and displacive facies; muddy towards top and base; erosion surface at 183.4 m | 6.7 | 188.8 |
| Mudstone, silty, grey at top, red below; zoned halite crystals | 0.5 | 189.3 |
| Mudstone, silty, mostly red with abundant displacive halite crystals | 1.5 | 190.8 |
| Halite rock, mostly displacive facies and apparently largely recrystallised; red and grey silty mudstone matrix; layered at intervals | 12.9 | 203.7 |

**Winsford No. 5 Borehole** (66 SW/1)
Surface level 58.5 m above OD
National Grid reference 6481 6453
Drilled 1971 for Cheshire County Council
Log by R. S. Arthurton

| | Thickness m | Depth m |
|---|---|---|
| No cores or samples | 170.7 | 170.7 |

MERCIA MUDSTONE GROUP

NORTHWICH HALITE
(Gamma ray log indicates top of salt at 166.42 m)

| | Thickness m | Depth m |
|---|---|---|
| Halite rock; displacive facies interspersed with layered; mostly grey silty mudstone matrix | 2.1 | 172.8 |
| Siltstone, mostly grey; cross-laminated in part; thin halite layer and scattered halite crystals near base | 1.8 | 174.6 |
| Halite rock; mostly displacive facies with red and grey silty mud matrix to 178.9 m and in basal 0.5 m; layered facies elsewhere with fairly pure halite | 8.6 | 183.2 |
| Siltstone and silty mudstone, part cross-laminated; scattered zoned halite crystals; halite vein | 0.6 | 183.8 |
| Halite rock; mostly layered facies; silty mudstone layer near base | 2.1 | 185.9 |

| | Thickness m | Depth m |
|---|---|---|
| Siltstone and silty mudstone, red and grey; load casts and cross-lamination at top; scattered halite crystals and 25 cm poorly layered halite near top; disturbed base with mudstone pellets | 1.9 | 187.8 |
| Halite rock, mostly displacive facies with red and grey silty mudstone matrix; interspersed layered halite rock | 7.1 | 194.9 |
| Mudstone, silty massive; scattered halite crystals | 0.6 | 195.5 |
| Halite rock, mostly displacive facies with grey or red mud matrix; a few thin units of layered halite; red silty mudstone layers in bottom metre | 5.3 | 200.8 |
| Mudstone, silty; massive at base, grey and laminated above; halite vein | 0.5 | 201.3 |
| Halite rock; mostly displacive facies with grey silty mudstone matrix; a few thin layers of silty mudstone; zoned halite crystals near top | 4.7 | 206.0 |

**Winsford No. 6 Borehole** (66 NW/13)
Surface level 60.8 m above OD
National Grid reference 6413 6512
Drilled 1970 for Cheshire County Council
Log by R. S. Arthurton (*see* Arthurton, 1980, pp. 52–53)

| | Thickness m | Depth m |
|---|---|---|
| No cores or samples | 32.0 | 32.0 |
| MERCIA MUDSTONE GROUP | | |
| Laminated red and grey mudstone and siltstone, alternating with blocky brownish red homogeneous silty mudstone. Laminated units thicker in lower part, thinner in upper part. The average thickness of 26 laminated/blocky units is 4.75 m; sporadic fibrous gypsum veins, associated with halite below 105.9 m | 123.7 | 155.7 |
| NORTHWICH HALITE | | |
| Solution residue, uncemented siltstone and mudstone breccia | 2.8 | 158.5 |
| Halite rock, displacive with red silty mudstone in upper half, layered below | 7.0 | 165.5 |
| Siltstone, red and grey; cross-laminated; halite crystals | 1.4 | 166.9 |
| Halite rock, displacive with grey silty mudstone matrix | 9.0 | 175.9 |
| Siltstone and silty mudstone with layers of halite | 5.4 | 181.3 |
| Halite rock, displacive and layered facies; three silty mudstone layers in upper part | 14.7 | 196.0 |
| Mudstone, silty, mostly massive | 0.6 | 196.6 |
| Halite rock, part layered part displacive; red and grey silty mudstone matrix | 16.8 | 213.4 |
| Siltstone, red with scattered halite crystals | 0.3 | 213.7 |
| Halite rock, mostly layered and fairly pure, some displacive facies | 4.5 | 218.2 |
| Mudstone, grey silty; veins of fibrous halite | 0.5 | 218.7 |
| Mudstone, silty, mainly red; laminated red and grey near base | 2.3 | 221.0 |
| Siltstone and silty mudstone (cores disturbed) | 5.0 | 226.0 |

**Worleston Borehole** (65 NE/1)
Surface level 45.8 m above OD
National Grid reference 6585 5749
Drilled 1958 for Geological Survey of Great Britain
Log by B. J. Taylor

| | Thickness m | Depth m |
|---|---|---|
| *Rock-bit drilling* | | |
| QUATERNARY | | |
| Made ground | 0.91 | 0.91 |
| Clay, red stony | 1.83 | 2.74 |
| Clay, red-brown stoneless | 10.98 | 13.72 |
| Clay, red stony | 1.82 | 15.54 |
| MERCIA MUDSTONE GROUP | | |
| Mudstone, red and grey with gypsum | 2.75 | 18.29 |
| *Core-drilling* | | |
| Mudstone, red-brown blocky facies with green spots and mottlings, alternating with laminated facies of red and grey mudstone, silty mudstone and some siltstone. The blocky facies predominates. Gypsum veins abundant throughout, associated with halite below 40 m. Desiccation cracks, turbulent bedding, penecontemporaneous deformation | 84.88 | 103.17 |
| Alternations of laminated and blocky facies, laminated facies predominates; and generally more arenaceous than above. Gypsum present mainly as thin primary layers or irregular masses. Desiccation cracks, penecontemporaneous deformation. Salt inclusions near base | 39.17 | 142.34 |
| Rock salt; much included mudstone | 7.6 | 149.94 |
| Mudstone and sandy mudstone interlaminated; saliferous | 2.16 | 152.10 |
| Rock salt, orange-brown; mudstone inclusions and beds of mudstone with bases at 161 m, 164 m, 172 m, 179 m | 28.65 | 180.75 |
| Rock salt; mainly of high purity; banded light and dark (orange and brown) at intervals | 17.47 | 198.22 |
| Mudstone, saliferous; beds of muddy salt with bases at 203 m, 206 m and 208 m | 12.75 | 210.97 |
| Rock salt, red-orange; mudstone inclusions | 2.39 | 213.36 |
| Mudstone very saliferous at top | 1.52 | 214.88 |
| Rock salt, mudstone inclusions at intervals | 7.93 | 22.81 |
| Rock salt, abundant mudstone inclusions | 3.58 | 226.39 |
| Mudstone, grey and brown saliferous | 0.56 | 226.95 |

# APPENDIX 2

## List of Geological Survey photographs

Copies of these photographs are deposited for reference in the libraries of the Geological Museum, Exhibition Road, London SW7 2DE and the British Geological Survey, Keyworth, Nottingham NG12 5GG. With the exception of A7414 which was taken in 1937 all the Series A photographs were taken in 1958–1959; the Series L photographs were taken in 1978. National Grid references to the viewpoints are all in 100-kilometre square SJ. The photographs may be supplied as black and white prints or lantern slides, and as colour prints or 2 × 2 colour transparancies, at a fixed tariff.

### Pleistocene and Recent

A7412   Disturbed glacial sand and gravel, sandpit, Northwich [652 742]

A9348   Glacial sands overlain by laminated clay and till. Beeston Castle sandpit [554 595]

A9349   Cross-bedded glacial sands. Beeston Castle sandpit [554 595]

L2192–6   Beeston Castle sandpit 20 years after A9348–9

A9393   Glacial sand-mound topography. View east from hill top just east of the railway arch (the dark arc) over the lane from Rangeway Bank to Brine's Brow [524 711]

A9394   Dissected glacial sand topography. Viewpoint 460 m east of Newpool Farm, Norley [576 715]

A9396   Canalised reach of river Weaver; floodplain and fluvio-glacial terrace. Viewpoint immediately east of Hartford Bridge [648 714]

A9397   Contact of till over glacial sands. Large disused sandpit on north-west side of Cuddington Brook north of Ravenscroft [591 717]

A9398   Shear plane in glacial sands. Sandpit east of Sandiway, about 640 m north-west of Earnshaw Grange [620 709]

A9399   Glacial sands. Sandpit east of Sandiway, about 640 m north-west of Earnshaw Grange [620 709]

A9400   Cross-bedding in glacial sands. Sandpit east of Sandiway, about 640 m north-west of Earnshaw Grange [620 710]

A9401   Contact of till on glacial sands. Large disused sandpit on north-west side of Cuddington Brook, north of Ravenscroft [591 717]

A9402   Cross-bedded glacial sands beneath till. Sandpit 410 m NNW of Ivy Cottage, Church Minshull [663 614]

A9403   Detail of cross-bedded glacial sands. Sandpit 410 m NNW of Ivy Cottage, Church Minshull [663 614]

A9404   Detail of cross-bedded glacial sands. Sandpit 410 m NNW of Ivy Cottage, Church Minshull [663 614]

A9405   Cross-bedded glacial sands beneath till. Over sandpit north side of Woodford Lane, 365 m west of Over crossroads [637 659]

A9406/L2207   Depression caused by salt solution. Marton Hole, 460 m west of Marton Hall [619 674]

A9407   A water-filled salt solution subsidence. Moulton, 275 m SSE of Hillside Farm [656 688]

A9408   Linear salt solution subsidence hollow. Winsford, 275 m NNW of Weaverwood Farm [663 638]

A9409/L2208   Hatchmere, a natural mere in a self-enclosed hollow in glacial sands [553 720]

A9415   Glacial drainage channel. Peckforton Gap, Bulkeley [531 554]

L2197   Recent landslip scars due to salt-solution subsidence alongside Bottom Flash, Winsford [658 649]

L2198   Bottom Flash, Winsford, a lake formed where the river Weaver crosses a salt-solution subsidence area [664 649]

L2199   Bottom Flash, Winsford; viewpoint 400 m west of L2198 [660 649]

L2200   Recent landslip scars due to salt-solution subsidence at Bottom Flash, Winsford [660 649]

L2201   Alternative view of L2200

L2202   'The Ocean' near Knight's Grange, Winsford, a salt-solution subsidence [636 672]

L2203   Alternative view of L2202

L2204   Alternative view of L2202

L2205   Bogart Brook, forming a lake (flash) where it crosses an area of salt-solution subsidence [631 681]

L2206   Alternative view of L2205

'L2209   Nunsmere, a lake after sand abstraction, not due to subsidence [588 690]

L2210   Alternative view of L2209

### Mercia Mudstone Group

A9367   Tarporley Siltstones with beds of 'marl-shale' and laminated sandstone. Quarry east of Lower Longley Farm, Kelsall [528 692]

A9368   Detail of Tarporley Siltstones with beds of 'marl-shale' and laminated sandstone. Quarry east of Lower Longley Farm, Kelsall [528 692]

A9369   Tarporley Siltstones with 'marly' partings. Road cutting on the Utkinton to Tarporley road, north of Holly Bank [550 643]

A9373   Footprints of *Chirotherium* on ripple-marked mudstone of the Tarporley Siltstones. Eaton Pumping Station No. 2 Borehole, depth 3.5 m [568 634]

### Sherwood Sandstone Group

A9347/L2189   Bold hill of Upper Mottled Sandstone capped by Helsby Sandstone. Beeston Castle Hill, viewed from the west [528 593]

A9350/L2186   Upper Mottled Sandstone, Beeston Castle Grotto [539 589]

A9351/L2187   Upper Mottled Sandstone, Beeston Castle Grotto [539 589]

A9352/L2188   Mineralised and discoloured Upper Mottled Sandstone, Beeston Castle Grotto [539 589]

A9353   Ribs of calcite in Upper Mottled Sandstone, Beeston Castle Grotto [539 589]

A9354/L2190   Helsby Sandstone. Outcrop in moat of Beeston Castle [538 593]

A9355   Upper Mottled Sandstone. Quarry immediately south of Burwardsley Hall Farm, Burwardsley [519 567]

A9358   Outcrop of Pebble Beds showing cross-bedded sandstone overlain by softer shingle-conglomerate. In Dunham-on-the-Hill village [472 727]

A9359   Cross-bedded Pebble Beds with band of shaly sandstone. Crag in the ravine of Bache Brook, just south of Bache Hall, Chester [402 680]

A9360/L2182   Outcrop of thickly bedded Pebble Beds at the Railway Inn Little Barrow [470 700]

A9361/L2183   Outcrop of thickly bedded Pebble Beds with few pebbles. Laneside cutting at entrance to the Rectory, Great Barrow [469 684]

A9362/L2184   Sandstone in Pebble Beds showing highly irregular bedding with very few pebbles. Laneside exposure near western entrance in Barrowmore Hospital [475 693]

A9363/L2185   Pebble-free sandstone in Pebble Beds showing anticlinal feature with 'balled-up' core. Laneside exposure near western entrance to Barrowmore Hospital [475 693]

A9364   Section of Pebble Beds dipping gently eastwards. Deep canal cutting beneath north wall of city, just east of Northgate, Chester [404 668]

A9365   Fault throwing Helsby Sandstone down east (to the right) against Upper Mottled Sandstone (left). Manley Quarry, in cutting for former railway siding, just east of the old road bridge over the siding [498 719]

A9366   Massive Helsby Sandstone. Trimmed face of Manley Quarry [478 719]

A9370   Topography made by Helsby Sandstone dipping gently eastwards. View northwards from near Lord's Well, Manley [515 728]

A9371   A view of escarpment of Helsby Sandstone north of Willington Corner. Looking north from a position north of Willington Hall [533 664]

A9372   Steep-sided ravine in Helsby Sandstone, running south-westwards from Roughlow Farm to Willington Corner [535 668]

A9395   Soft Upper Mottled Sandstone worked for sand. Ship Canal Sand Quarry, on north side of railway east of former Mouldsworth Station [516 709]

A9412   Helsby Sandstone crags surmounted by Beeston Castle. View from the west [528 593]

A9413   Prominent west-facing scarp formed by Helsby Sandstone capping the upper part of the Upper Mottled Sandstone. Burwardsley Hill viewed from the south-west [504 554]

A9414   Prominent north-west facing scarp formed by Helsby Sandstone underlain by Upper Mottled Sandstone. Viewed south-westwards from Beeston village towards Peckforton Castle [541 587]

# INDEX

## BRITISH GEOLOGICAL SURVEY

Keyworth, Nottinghamshire NG12 5GG

Murchison House, West Mains Road,
Edinburgh EH9 3LA

The full range of Survey publications is available through the Sales Desks at Keyworth and Murchison House. Selected items are stocked by the Geological Museum Bookshop, Exhibition Road, London SW7 2DE; all other items may be obtained through the BGS London Information Office in the Geological Museum. All the books are listed in HMSO's Sectional List 45. Maps are listed in the BGS Map Catalogue and Ordnance Survey's Trade Catalogue. They can be bought from Ordnance Survey Agents as well as from BGS.

*On 1 January 1984 the Institute of Geological Sciences was renamed the British Geological Survey. It continues to carry out the geological survey of Great Britain and Northern Ireland (the latter as an agency service for the government of Northern Ireland), and of the surrounding continental shelf, as well as its basic research projects. It also undertakes programmes of British technical aid in geology in developing countries as arranged by the Overseas Development Administration.*

*The British Geological Survey is a component body of the Natural Environment Research Council.*

## HER MAJESTY'S STATIONERY OFFICE

HMSO publications are available from:

**HMSO Publications Centre**
(Mail and telephone orders)
PO Box 276, London SW8 5DT
Telephone orders (01) 622 3316
General enquiries (01) 211 5656

**HMSO Bookshops**
49 High Holborn, London WC1V 6HB
 (01) 211 5656 (Counter service only)
258 Broad Street, Birmingham B1 2HE
 (021) 643 3757
Southey House, 33 Wine Street, Bristol BS1 2BQ
 (0272) 24306/24307
9 Princess Street, Manchester M60 8AS
 (061) 834 7201
80 Chichester Street, Belfast BT1 4JY
 (0232) 238451
13a Castle Street, Edinburgh EH2 3AR
 (031) 225 6333

**HMSO's Accredited Agents**
(see Yellow Pages)

*And through good booksellers*